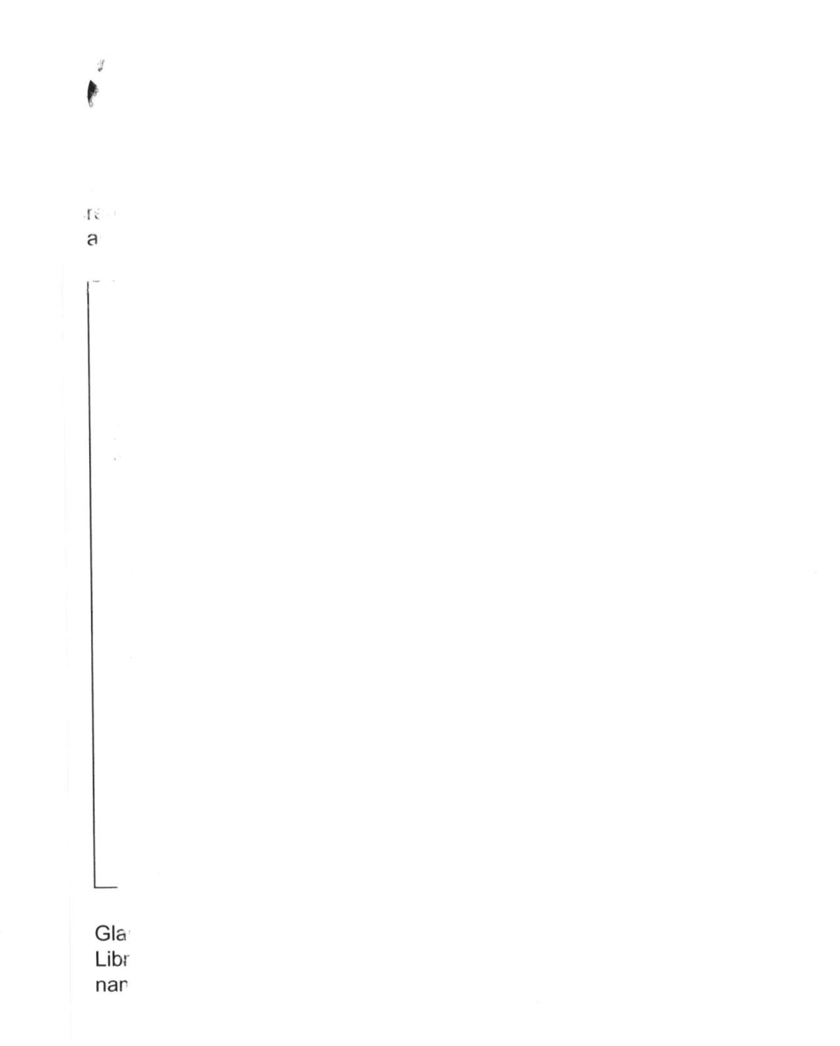

Move Fast and Break Things

Also by Jonathan Taplin

Outlaw Blues: Adventures in the Counter-Culture Wars

Move Fast and Break Things

How Facebook, Google and Amazon have cornered culture and what it means for all of us

Jonathan Taplin

MACMILLAN

First published 2017 by Little, Brown and Company,
a division of Hachette Book Group, Inc., New York

First published in the UK 2017 by Macmillan
an imprint of Pan Macmillan
20 New Wharf Road, London N1 9RR
Associated companies throughout the world
www.panmacmillan.com

ISBN 978-1-5098-4769-3

1 3 5 7 9 8 6 4 2

A CIP catalogue record for this book is available from the British Library.

Printed and bound by CPI Group (UK) Ltd, Croydon, CR0 4YY

For Maggie

Move fast and break things. Unless you are
breaking stuff, you aren't moving fast enough.

— Mark Zuckerberg

Contents

Move Fast and Break Things

Introduction

I thought I was going to write the story of a culture war. On one side were a few libertarian Internet billionaires – the people who brought you Google, Amazon, and Facebook – and on the other side were the musicians, journalists, photographers, authors, and filmmakers who were trying to figure out how to continue to make a living in the digital age. I have spent much of my life producing music and movies for artists such as Bob Dylan and The Band, George Harrison, and Martin Scorsese, to name a few, and the future of the media in which I worked – not to mention the role of the artist in our society – is important to me. I was lucky enough to start out at a time when an artist could make a decent living by making music or a movie, and as a partner in this work, I succeeded, too. But those days are over. Since 1995 – the last time I produced a movie (*To Die For*) – the digital distribution of most popular forms of art has reinforced the popularity of a small group of artists and cast almost all others into shadow. To be a young musician, filmmaker, or journalist today is to seriously contemplate the prospect of entering a profession that the digital age has eroded beyond recognition.

The deeper you delve into the reasons artists are struggling in the digital age, the more you see that Internet monopolies are at the heart of the problem and that it is no longer a problem just for artists. The Web has become critical to all of our lives as well as the world economy, and yet the decisions on how it is designed have never been voted upon by anyone. Those decisions were made by engineers and libertarian executives at Google, Facebook, and Amazon (plus a few others) and imposed upon the public with no regulatory scrutiny. We have also seen a plethora of new platforms like Uber, Airbnb and Twitter, which are operating in an unregulated environment that radically changes the world we inhabit. The result is what President Obama calls "a Wild West" world without privacy or security that leaves every citizen vulnerable to criminal, corporate, and government intrusion. As Obama wrote in *The Economist*, "a capitalism shaped by the few and unaccountable to the many is a threat to all."

The Internet is changing our democracy, too: in Twitter, Donald Trump found the perfect vehicle for his narcissistic personality, allowing him to strike out at all his perceived tormentors. And Facebook (the primary news source for 44 percent of Americans) was equally responsible for the Trump victory, according to Ed Wasserman, the dean of the University of California, Berkeley, Graduate School of Journalism: "Trump was able to get his message out [on Facebook] in a way that was vastly influential without undergoing the usual kinds of quality checks that we associate with reaching the mass public." Facebook was flooded with fake news stories, and *BuzzFeed* reported that "in the final three months of the US presidential campaign, the top-performing fake election news stories on Facebook generated more engagement than the top stories from major news outlets such as the *New York Times, Washington Post, Huffington Post,* NBC News, and others." As Ian Bremmer, president of the Eurasia Group told the *New York Times,* "If it wasn't for social media, I don't see Trump winning."

But the libertarians who control some of the major Internet firms do not really believe in democracy. The men who lead these monopolies believe in an oligarchy in which only the brightest and richest get to determine our future. Peter Thiel, the first outside investor in Facebook and cofounder of PayPal, thinks the major problem of American society is its "unthinking demos": the democratic public that constrains capitalism. Thiel told *Wall Street Journal* columnist Holman W. Jenkins that only 2 percent of the populace – the scientists, entrepreneurs, and venture capitalists – understand what is going on and "the other 98 percent don't know anything."

What I mistook as only a culture war is an economic war. It is likely only a preview of capitalism in the digital age. *The Economist,* in a special issue on monopoly capitalism entitled "Winners Take All," editorialized that perhaps "firms are abusing monopoly positions, or using lobbying to stifle competition. The game may indeed be rigged." They went on to suggest that what was needed was a major reform effort that "would involve more active, albeit cruder, antitrust actions. It would start a more serious conversation about whether it makes sense to have most of the country's data in the hands of a few very large firms. It would revisit the entire issue of corporate lobbying, which has become a key mechanism by which incumbent firms protect themselves." Monopoly, control of our data, and corporate lobbying are at the heart of this story of the battle between creative artists and the Internet giants, but we need to understand that every one of us will stand in the shoes of the artist before long. Musicians and authors were at the barricades first because their industries were the first to be digitized. But as the venture capitalist Marc Andreessen has said, "Software is eating the world," and soon the technologists will be coming for your job, too, just as they will continue to come for more of your personal data.

The rise of the digital giants is directly connected to the fall of the creative industries in our country. I would put the date of the real rise of digital monopolies at August of 2004, when Google raised $1.67 billion in its initial public offering. In December of 2004, Google's share of the search-engine market was only 35 percent. Yahoo's was 32 percent, and MSN was at 16 percent. Today, Google's market share is 88 percent in the United States and almost 91 percent in Europe. In 2004 Amazon had net sales revenue of $6.9 billion. In 2015 its net sales revenue was $107 billion, and it now controls 65 percent of all online new book sales, whether print or digital. In those eleven years, a massive reallocation of revenue – perhaps $50 billion per year—has taken place, in which economic value has moved from the creators of content to the owners of monopoly platforms.

Since 2000, global recorded music revenues have fallen from $27.3 billion to $10.4 billion per year according to the IFPI. Home-video revenue has fallen from $21.6 billion in 2006 to $18 billion in 2014. US newspaper advertising revenue has fallen from $65.8 billion in 2000 to $23.6 billion in 2014. Between 2007 and 2013 UK newspaper ad spend fell from £4.7 billion to £2.6 billion. The Pointer Institute estimates that "Facebook has sucked well over $1 billion out of print advertising budgets for US newspapers" in 2016. While book publishing revenues have remained flat, this is mostly because an increase in children's book sales has made up for the almost 30 percent fall in sales of adult titles. During that same period (2003–2015), Google revenue grew from $1.5 billion to $74.5 billion. According to *Adweek*, Google, as of 2016, is the largest media company in the world, collecting "$60 billion in U.S. ad spend—a figure 166 percent larger than [the] No. 2 ranking ... Walt Disney Company." Google's dominance in online advertising means that its revenue dwarfs the ad revenue collected by a TV giant like Walt Disney, which controls ABC, ESPN, and the Disney Channel. And because Google has such a large share of advertising revenue, global brands are paying the company (and Facebook) a premium, which of course is passed on to consumers in the form of higher prices.

The astonishing and precipitous decline in revenue paid to content creators has nothing to do with the idea that people are listening to less music, reading less, or watching fewer movies and TV shows. In fact all surveys point to the opposite — the top-searched Google items are all about entertainment categories. It is not a coincidence that the rise of digital monopolies has led to the fall of content revenues. The two are inextricably linked.

The five largest firms in the world (in terms of market capitalization) are Apple, Google (now referred to as Alphabet), Microsoft, Amazon, and Facebook. It is hard to grasp just how large a role these five tech giants play in our economy, but look at this comparison of the world's largest companies in 2006 versus today.

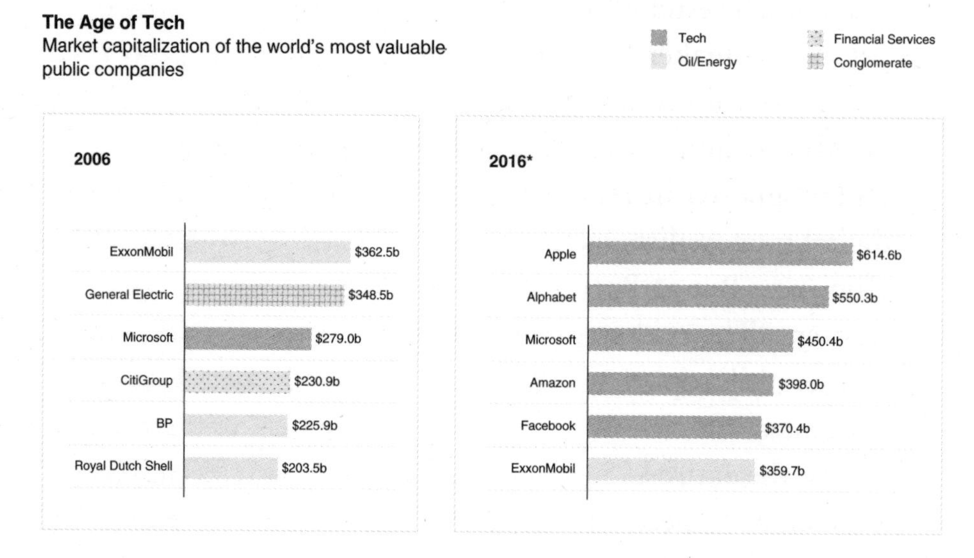

But the change is more profound than just market capitalization. Not since the turn of the twentieth century, when Theodore Roosevelt took on the monopolies of John D. Rockefeller and J. P. Morgan, has the country faced such a concentration of wealth and power. Peter Orszag and Jason Furman, economic advisers to President Obama, have argued that the fortunes created by the digital revolution may have done more to increase economic inequality than almost any other factor. Despite Marc Andreessen's and Peter Thiel's belief that the outsize gains of tech billionaires are the result of a genius entrepreneur culture, inequality at this scale is a choice — the result of the laws and taxes that we as a society choose to establish. Contrary to what techno-determinists want us to believe, inequality is not the inevitable by-product of technology and globalization or even the lopsided distribution of genius. It is a direct result of the fact that since the rise of the Internet, policy makers have acted as if the rules that apply to the rest of the economy do not apply to Internet monopolies. Taxes, antitrust regulation, intellectual property law — all are ignored in regulating the Internet industries. The digital monopolists have argued for free rein in pursuit of efficiency. But as Barry Lynn and Phillip Longman have written, "The evidence is close to irrefutable that adoption of this philosophy of 'efficiency' unleashed a process of concentration that over the last generation has remade almost the entire U.S. economy, and is now disrupting our democracy." Clearly the increasing concentration of market share in the hands of a few US corporations ranges far beyond the tech sector, as Senator Elizabeth Warren pointed out in a speech she gave in June of 2016:

> In the last decade, the number of major U.S. airlines has dropped from nine to four. The four that are left standing — American, Delta, United, and Southwest — control over 80% of all domestic airline seats in the country.... A

> handful of health insurance giants — including Anthem, Blue Cross Blue Shield, United Healthcare, Aetna, and Cigna — control over 83% of the country's health insurance market.... Three drug stores — CVS, Walgreens, and Rite Aid — control 99% of the drug stores in the country. Four companies control nearly 85% of the U.S. beef market, and three companies produce almost half of all chicken.

While acknowledging the problem of increased concentration of power in the hands of a few giants across all global industries, I am going to focus here on the industry I have spent my life in — media and communication. In this world the relentless pursuit of efficiency leads Google, Amazon, and Facebook to treat all media as a commodity, the real value of which lies in the gigabytes of personal data scraped from your profile as you peruse the latest music video, news article, or listicle. But the people who make the work that drives the Internet are critical to our understanding of who we are as a civilization.

Throughout history the artist has pointed out the injustices of society. The philosopher Herbert Marcuse wrote that the role of art in a society is "in its refusal *to forget what can be*" (the italics are mine). The history of art is the history of subversion, of a person like Galileo saying that everything you know is wrong. The transcendentalism of Emerson and Thoreau in the 1830s was the first "great refusal" — the refusal to accept slavery and American imperialism — which thirty years later produced Lincoln's Emancipation Proclamation. This pattern of artists being in the vanguard of progress has been repeated numerous times in our history (and the history of many nations), and while rebel artists have always had to deal with "the suits" who control the channels of distribution, the coming of the Internet monopolies has placed all of us in the vulnerable position that the artist alone used to occupy. The concentration of profits in the making of art and news has made more than just artists and journalists vulnerable: it has made all those who seek to profit from the free exchange of ideas and culture vulnerable to the power of a small group of powerful patrons. Even Google's chairman, Eric Schmidt (and his coauthor Jared Cohen), acknowledged this when they wrote, "We believe that modern technology platforms, such as Google, Facebook, Amazon and Apple, are even more powerful than most people realize, and our future world will be profoundly altered by their adoption and successfulness in societies everywhere."

Martin Luther King Jr. delivered a sermon at the National Cathedral in Washington, DC, less than a week before his assassination, in 1968. He asserted that although we were embarking on a technological revolution, many were blind to the changes it would bring, and without some sort of moral framework we would have what he once referred to as "guided missiles and misguided men." He said:

> One of the great liabilities of life is that all too many people find themselves living amid a great period of social change, and yet they fail to develop the new attitudes, the new

> mental responses, that the new situation demands. They end up sleeping through a revolution.

Think about the fact that Dr. King, with all the battles he believed still lay before him – civil rights, the Vietnam War, poverty – sought to focus our attention on the role technology might play in our future. King had no way of envisioning the addictive nature of the Internet, a place where we would be willing to share our most intimate secrets with a faceless corporation whose business model is to get into our heads and harvest our attention. And as any parent of a teenager who sleeps with a smartphone will agree, one hardly needs to be awake to interface with Google or Facebook. We continue to surrender more of our private lives believing in the myth of convenience bequeathed to us by benign corporations. As Kevin Kelly, the founding editor of *Wired,* remarked, "Everything will be tracked, monitored, sensored, and imaged, and people will go along with it because 'vanity trumps privacy,' as already proved on Facebook. Wherever attention flows, money will follow." But Kelly, one of the original techno-determinists, may be wrong. Speaking at the Black Hat USA cybersecurity conference in 2015, longtime tech security guru Dan Kaminsky said, "Half of all Americans are backing away from the net due to fears regarding security and privacy. We need to go ahead and get the Internet fixed or risk losing this engine of beauty."

People like Google CEO Larry Page, Facebook's Mark Zuckerberg, PayPal founder Peter Thiel, and Sean Parker of Napster and Facebook fame are among the richest men in the world, with ambitions so outsize that they are the stuff of fiction: Dave Eggers's *The Circle* and Don DeLillo's *Zero K* are populated with tech billionaires inventing technology that will enable people to live forever. But this scenario is happening in real life. Peter Thiel, Larry Page, and others are investing hundreds of millions of dollars in research to "end human aging" and to merge human consciousness into their all-powerful networks. As George Packer, writing in the *New Yorker,* put it, "In Thiel's techno-utopia a few thousand Americans might own robot-driven cars and live to a hundred and fifty, while millions of others lose their jobs to computers that are far smarter than they are, then perish at sixty." Surprisingly, we have just passed through a presidential election campaign where these issues of the future were never even raised.

Modernity was constructed on the idea that individuals are meant to be the ones who decide their own fates, particularly as voters and consumers. But that's not what the techno-determinist future holds. As former Google "design ethicist" Tristan Harris wrote, "If you control the menu, you control the choices." We cede our freedom to choose by giving networks like Google and Facebook control of the menu. How the mysterious black-box algorithms of Google, Facebook, and Amazon determine the menu can never be known by anyone outside those companies. The former editor of the *Guardian* Alan Rusbridger told a *Financial Times* conference in September of 2016 that Facebook had "sucked up $27 million" of the paper's projected digital advertising revenue in the previous year. "They [Facebook] are taking all the money," he noted, because "they have algorithms we don't understand, which are a filter between what we do and how people receive it." As more of our lives become digital, these new algorithmic gods will assume more power over us.

This is not the first moral crisis brought on by the techno-determinist point of view. At the end of World War II, in the shadow of the mushroom clouds over Hiroshima and Nagasaki, Christian intellectuals such as Reinhold Niebuhr worried that we were "winning the war, but losing the peace." They thought that if technocrats were credited with winning the war, then those technocrats would manage the postwar world. Niebuhr was prescient on this point, and President Obama acknowledged the dichotomy in his 2016 speech at the Hiroshima Peace Memorial: "Technological progress without an equivalent progress in human institutions can doom us. The scientific revolution that led to the splitting of an atom requires a moral revolution as well." But as one travels through America's Rust Belt cities, where the forces of technology have destroyed jobs, one sees signs of real suffering — high rates of addiction and suicide and shortened life expectancies. What is the technology fix for these cancers of the spirit? The answer escapes me. Or think about the people — celebrities and others — who are subjected to anonymous hate-filled trolls on Twitter. What is the technology solution to this problem?

For all the outrage generated by Edward Snowden over National Security Agency spying, the average citizen has voluntarily (though unknowingly) turned over to Google and Facebook far more personal information than the government will ever have. And even if we are aware that Google's and Facebook's primary business is "surveillance marketing" – selling our personal information to advertisers for billions of dollars – we somehow trust that they will not exploit this information in ways that might harm us. "Google policy is to get right up to the creepy line," Eric Schmidt once told *The Atlantic,* "and not cross it," a debatable statement at best. As Snowden revealed, Google and Facebook are willing to turn over customer data to the NSA. Now imagine Google under the management of, say, Jeffrey Skilling of Enron fame. How easy might it be to cross the "creepy line."

Consider, for example, the ability of Google and Facebook to tweak their algorithms and thereby influence the choice of news stories you see. In 2014, a study led by Robert Epstein, a psychologist at the American Institute for Behavioral Research and Technology, analyzed the extent to which political candidates' Google search rankings could influence voters. Epstein noted, "We estimate, based on win margins in national elections around the world that Google could determine the outcome of upwards of 25 percent of all national elections." And research by Professor Jonathan Albright has shown that right-wing websites have manipulated Google's algorithm to autofill the query "Are Jews..." with the word *evil* as the top choice. He noted, "The right has colonized the digital space around these subjects—Muslims, women, Jews, the Holocaust, black people—far more effectively than the liberal left."

Google, Amazon, and Facebook are classic "rent-seeking" enterprises. The *New York Times* columnist Adam Davidson explains the concept:

> In economics, a "rent" is money you make because you control something scarce and desirable, whether it's an oil field or a monopolistic position in a market.... The left, right and center of the economics profession all agree that reducing rent-seeking behavior, and improving overall growth, is essential if we want to "make America great again."

Google and Facebook each have more than one billion customers while Amazon has 350 million. They all take their rent off the top, whether through direct payment or advertising subsidy. This rise of the new monopolies happened relatively quickly, so in a way economists and politicians do not fully understand how different monopoly capitalism is from the idealized Adam Smith capitalism that is still taught in economics 101. To begin with, monopolies are price makers, not price takers. As the economist Paul Krugman wrote, "Don't tell me that Amazon is giving consumers what they want, or that it has earned its position. What matters is whether it has too much power, and is abusing that power. Well, it does, and it is."

But the real effect of the fact that global business is tending toward more market-share concentration in all sectors is that corporate profits have been rising and wages stagnating since the 1970s. In a business landscape with high concentration in all sectors, the declining fortunes of the average worker mirror the predicament of the new musician, filmmaker, or journalist. If "data is the new oil" then many businesses will have to learn how to align themselves to Google, Amazon, and Facebook — the last surviving institutions that will make it possible to earn a living.

These companies did not arrive at their dominant position solely because of the brilliance of their founders, even though the business press would have you believe so. Their monopolies are examples of the effects a political theory called libertarianism, based on the work of economist Milton Friedman and philosopher Ayn Rand, which quite simply posits that government is usually wrong and the market is always right. Strikingly, the Internet was created with government funding and built on the principles of decentralization — principles we need to find our way back to if we are to overcome the power of corporate monopolies in the digital age.

Since 2010 I have run the Annenberg Innovation Lab at the University of Southern California, where I have been lucky enough to work with many of the pioneers of the Internet, including Tim Berners-Lee, Vint Cerf, and John Seely Brown. I was also the founder of one of the first streaming-video-on-demand companies, Intertainer, which deployed high-quality video over the Internet ten years before YouTube went online. I am a committed believer in the power of technology. I have used Internet tools such as my blog on *Medium* to work out some of the ideas in this book. But I'm not sure technology can solve what is primarily a values problem. How do artists derive monetary value from their work, and how do we as a society value art in the digital age? How do we create a sustainable culture that elevates our lives, our spirits, our souls — as did Louis Armstrong and Walt Whitman and Bob Dylan and Stanley Kubrick? You will see that I believe in the power of rock and roll, great writing, and breakthrough movies to change lives. So while I may paint a grim picture of our contemporary digital culture, I hope to show you a way for artists and citizens to help recapture the vision of the Internet pioneers in what I call a Digital Renaissance. Like the historical Renaissance, this one will begin with acts of resistance. This has already started with a revolt against YouTube by musicians, who were first to feel the effects of digitization. It is now spreading to journalists, filmmakers, and even politicians, including Senator Elizabeth Warren. The TV producer Kurt Sutter (*Sons of Anarchy*) spoke for many when he said, "Google spends millions of dollars every year fronting a campaign to crush the rights of creatives." In later chapters I will lay out both the breadth of the resistance and some of my own ideas to solve this problem.

But first we need to understand how we got here.

CHAPTER ONE

The Great Disruption

"Don't Be Evil"
— Google motto

1.

The beginnings of the technical and social revolution that Martin Luther King referenced in his 1968 sermon at the National Cathedral were under way even as he was speaking. The revolution began in the moral precepts of the counterculture: decentralize control and harmonize people. The earliest computer networks — like the Whole Earth 'Lectronic Link (WELL), organized by Stewart Brand, the founder of *The Whole Earth Catalog* — grew directly out of 1960s counterculture, and were an attempt to give the new commune movement "access to tools" that could foster a non-corporate way of earning a living. Brand had helped novelist Ken Kesey organize the Acid Tests — epic be-ins where thousands of hippies ingested LSD and danced to the music of a new band, the Grateful Dead. Steve Jobs, founder of Apple Computer, Inc., dropped acid as well. "Jobs explained," wrote John Markoff in his book *What the Dormouse Said*, "that he still believed that taking LSD was one of the two or three most important things he had done in his life, and he said he felt that because people he knew well had not tried psychedelics, there were things about him they couldn't understand." Brand, Kesey, and Jobs envisioned a new kind of network that was truly "bottom-up." But our hopes that this new kind of network would overthrow political hierarchies and decrease inequality have turned out to be pipe dreams, the fantasies of digital utopians. A *New York Times* article on a 2016 World Bank report noted, "Internet innovations stand to widen inequalities and even hasten the hollowing out of middle-class employment." How did something so promising go so wrong? As the MIT researcher and early Internet pioneer Ethan Zuckerman wrote, "It's obvious now that what we did was a fiasco, so let me remind you that what we wanted to do was something brave and noble."

The original mission of the Internet was hijacked by a small group of right-wing radicals to whom the ideas of democracy and decentralization were anathema. By the late 1980s, starting with eventual PayPal founder Peter Thiel's class at Stanford University, the dominant philosophy of Silicon Valley would be based far more heavily on the radical libertarian ideology of Ayn Rand than the commune-based principles of Ken Kesey and Stewart Brand. Thiel, who was also an early investor in Facebook and is the godfather of what he proudly calls the PayPal Mafia, which currently rules Silicon Valley, has been clear about his credo, stating, "I no longer believe that freedom and democracy are compatible." More important, Thiel says that if you want to create and capture lasting value, you should look to build a monopoly. Three of the companies that have played the largest role in imperiling artists are clear monopolies. Google has an 88 percent market share in online searches and search advertising. Google's Android mobile operating system has an 80 percent global market share in its category. Amazon has a 70 percent market share in ebook sales. Facebook has a 77 percent market share in mobile social media. The fourth firm, Apple, is not a monopoly because its main hardware business has many competitors. Apple has a role to play in this story, but I will focus on the three digital monopolies that have done the most to alter the relationship between artists and those who support their work.

Perhaps not since the day of John D. Rockefeller's Standard Oil has a single company dominated a market as Google does. While Google and Facebook use their market power to extract monopoly rents from advertisers that are often 20 percent higher than market price, Amazon uses its monopsony (a market structure in which only one buyer interacts with many would-be sellers) to force authors, publishers, and booksellers to lower their prices, putting many of them out of business. This was not the decentralization that the founders of the Internet imagined, but ironically it was the very design of the Internet, with a set of global standards that allowed for massive scale, that led to the winner-takes-all economy of the Internet age.

In another time, Google, Facebook, and Amazon

would have been subject to government strictures and might be half the size they are now because much of their growth has been through acquisition, which would have been prevented by strict antitrust-law enforcement. Running for president in 1912, Woodrow Wilson said, "If monopoly persists, monopoly will always sit at the helm of the government. What we have to determine now is whether we are big enough, whether we are men enough, whether we are free enough, to take possession again of the government which is our own." But the antiregulatory religion pushed by libertarian think tanks since the Reagan era has gutted antitrust enforcement efforts in both Republican and Democratic administrations. As former secretary of labor Robert Reich wrote in 2015, "Big Tech has been almost immune to serious antitrust scrutiny, even though the largest tech companies have more market power than ever. Maybe that's because they've accumulated so much political power."

Even the investment bank Goldman Sachs finds itself perplexed by the extraordinary margins these monopolies are generating. In the standard model of capitalism, super-high-profit businesses encourage new competitors to enter the market, eventually forcing profits to normalize (economists call this mean reversion). A Goldman Sachs report says, "We are always wary of guiding for mean reversion. But, if we are wrong and high margins manage to endure for the next few years (particularly when global demand growth is below trend), there are broader questions to be asked about the efficacy of capitalism." Coming from the premier investment bank on Wall Street, this idea that capitalism is no longer working is quite an amazing statement.

This gets to the heart of our inquiry. Popular entertainment has had a "free" model, sponsored by advertising, since the dawn of radio in 1920. But companies like the National Broadcasting Company (NBC) and the Columbia Broadcasting System (CBS), which pioneered commercial broadcasting, always reinvested a large portion of their profits in the creation of content. Google, YouTube, and Facebook, by contrast, invest nothing in the creation of content – it's all "user-generated," even though much of it is professionally produced and appropriated by users. Even today CBS's profit margin is 11 percent compared to Google's 22 percent. That 11 percent advantage for Google represents potentially $8 billion not invested in content production.

The Economist reported in 2016 that "an American firm that was very profitable in 2003 (one with post-tax returns on capital of 15–25%, excluding goodwill) had an 83% chance of still being very profitable in 2013; the same was true for firms with returns of over 25%, according to McKinsey, a consulting firm. In the previous decade the odds were about 50%. The obvious conclusion is that the American economy is too cozy for incumbents." The power of incumbency also limits the number of new start-up businesses, which have historically been the source of American job growth. New research by MIT economists Scott Stern and Jorge Guzman shows that "even as the number of new ideas and potential for innovation is increasing, there seems to be a reduction in the ability of companies to scale in a meaningful and systematic way. It has become increasingly advantageous to be an incumbent, and less advantageous to be a new entrant." Even when a new entrant like Instagram, Twitch or WhatsApp enters the market, they are quickly acquired by the monopoly firms.

In 2013, Balaji Srinivasan, now a partner at the venture-capital company Andreessen Horowitz, claimed that Silicon Valley was becoming more powerful than Wall Street and the US government. He noted, "We want to show what a society run by Silicon Valley would look like. That's where 'exit' comes in. . . . It basically means: build an opt-in society, ultimately outside the US, run by technology. And this is actually where the Valley is going. This is where we're going over the next ten years. . . . [Google cofounder] Larry Page, for example, wants to set aside a part of the world for unregulated experimentation." This is not just a libertarian fantasy. This is where Peter Thiel and Larry Page want to take the world. Thiel has financially supported an idea called seasteading, which is the concept of creating permanent artificial islands, called seasteads, outside the territory claimed by any government. These cloud businesses could thereby escape taxation and regulation. Page has financed extensive research on privately owned city-states. But President Obama has cautioned the Silicon Valley leaders, saying:

> Sometimes we get, I think, in the tech community, the entrepreneurial community, the sense of — we just have to blow up the system, or create this parallel society and culture because government is inherently wrecked. No, it's not inherently wrecked; it's just government has to

> care for, for example, veterans who come home. That's not on your balance sheet, that's on our collective balance sheet, because we have a sacred duty to take care of those veterans. And that's hard and it's messy, and we're building up legacy systems that we can't just blow up.

But this sense of shared social responsibility is not part of the libertarian creed, which in many respects is antidemocratic. As Ben Tarnoff, writing in the *Guardian* noted, one of the reasons Peter Thiel was drawn to Donald Trump's authoritarian candidacy was that "he would discipline what Thiel calls 'the unthinking demos': the democratic public that constrains capitalism."

But for now there are few constraints on Tech capitalism. The monopoly profits of this new era have been very, very good to a few men. The Forbes 400 list, which ranks American wealth, places Bill Gates, Larry Ellison, Larry Page, Jeff Bezos, Sergey Brin, and Mark Zuckerberg in the top ten. The Silicon Valley venture capitalist Paul Graham (CEO of Y Combinator, the largest start-up incubator in the US), in a 2016 blog post, was quite open about celebrating income inequality. He wrote, "I've become an expert on how to increase economic inequality, and I've spent the past decade working hard to do it. Not just by helping the 2500 founders YC has funded. I've also written essays encouraging people to increase economic inequality and giving them detailed instructions showing how."

If tech billionaires have achieved political and economic power unseen since the Gilded Age, they have also achieved cultural power. David Nasaw, Andrew Carnegie's biographer, has said, "Carnegie could never have imagined the kind of power Zuckerberg has. Politics today is less relevant than it has ever been in our entire history. These CEOs are more powerful than they've ever been. The driving force of social change today is no longer government at all." And the libertarian ideology of Silicon Valley has seeped its way into pop culture. As the New York Times film critic A. O. Scott pointed out, in our current fascination with superhero movies,

> the genre's default ideology is a variant of the masters-of-the-universe libertarianism that energizes some of the most vocal sectors of the American ruling class. The supermen are doing good, and they know what's good for us, and they have never needed pusillanimous institutions—the cops, the press, the government—to tell them what to do. What they need is the support and gratitude of the masses, and when they don't get that affirmation, they can get a little sulky.

Mark Zuckerberg and Larry Page want our gratitude because they would have us believe they've delivered a period of unprecedented innovation that will inevitably improve not just their lives but also the lives of all the earth's citizens. But is this true? Statistics from the Organisation for Economic Co-operation and Development (OECD) tell another tale: economic growth is dramatically slowing while inequality in developed countries is increasing. Unlike the years of more than 6 percent growth spurred by twentieth-century innovation cycles (electricity, communication, transportation), the digital revolution is delivering less than 2 percent growth and increasing inequality in the developed world. As economist Paul Krugman notes, reviewing Robert Gordon's *The Rise and Fall of American Growth: The U.S. Standard of Living Since the Civil War,* "Gordon suggests that the future is all too likely to be marked by stagnant living standards for most Americans, because the effects of slowing technological progress will be reinforced by a set of 'headwinds': rising inequality, a plateau in education levels, an aging population and more." If in fact profits are accruing to increasingly dominant tech-industry monopolies, a process that is eliminating middle-class jobs (think robots and self-driving truck fleets), we can see that the techno-determinist path will ultimately lead to deep social unrest.

2.

We are trapped inside the libertarian economic and personal theories Milton Friedman and Ayn Rand created in the 1950s. For corporations, Friedman decreed, "there is one and only one social responsibility of business – to increase its profits." Rand told individuals that "achievement of your happiness is the only moral purpose of your life." As recently as the late 1970s these theories were viewed as the crackpot musings of reactionaries. A review of Rand's essay collection *Capitalism: The Unknown Ideal* in the *New Republic* simply referred to Rand as "Top Bee in the communal bonnet, buzzing the loudest and zaniest throughout this all but incredible book." But since the election of Ronald Reagan and Margaret Thatcher, these libertarian principles have won the battle of ideas. Since then, notions that the state should regulate the free market have been out of favor in both the US and UK. It may be that the Great Recession of 2008 led many to realize that this philosophy is a dead end for both culture and politics, but we seem to be lacking the political and cultural will to direct society onto a new path. Nobel Prize–winning economist Joseph Stiglitz thinks we need to rethink the laissez-faire economics of Rand and Friedman: "If markets are based on exploitation, the rationale for laissez-faire disappears. Indeed, in that case, the battle against entrenched power is not only a battle for democracy; it is also a battle for efficiency and shared prosperity."

It is important at the outset that we not assume that this technological revolution we are living through has but one inevitable outcome. I am reminded of the slogan British prime minister Margaret Thatcher used to use when talking about her program of deregulation and tax cuts for the rich: "TINA—There is no alternative." But history is made by humans, not by corporations or machines. Digital-age robber barons tell us that everything is different now and that they deserve to win because they are smart enough to throw away conventional wisdom and embrace disruption. But a culture and its art are not like an old flip phone—to be thrown in the trash as soon as it has been "disrupted" by the Next Big Thing. Culture thrives on continuity. As Pete Seeger once said, "Every songwriter is just a link in the chain."

A writer I greatly admire, the late Gabriel García Márquez, embodied, for me, the role of the artist in society. His life and work were marked by the refusal to believe that we are incapable of creating a more just world. Utopias are out of favor now. Yet García Márquez never gave up believing in the transformative power of words to conjure magic and seize the imagination. He also taught us the importance of regionalism. In a commercial culture of sameness, where you can stroll through a mall in Shanghai and forget that you're not in Los Angeles, García Márquez's work was distinctly Latin American, as unique as the songs of Gilberto Gil or the cinema of Alejandro González Iñárritu. In a culture like ours, which has long advocated a "melting pot" philosophy that papers over differences, it is valuable to recognize that allowing our dissimilarities to act as barriers is not the same as appreciating the things that make each culture unique, situated in time and space and connected to a particular people. What's more, young artists need to have the sense of history that García Márquez celebrated when he said, "I cannot imagine how anyone could even think of writing a novel without having at least a vague idea of the 10,000 years of literature that have gone before." Cultural amnesia only leads to cultural death. If the only university students who receive state help are computer engineers, we as a culture will lose something.

But Google, YouTube, and Facebook treat cultural objects as commodities — click bait. The scholar James Delong suggests that Google's principal mission is to commoditize the world's media:

> In most circumstances, the commoditizer's goal is restrained by knowledge that enough money must be left in the system

> to support the creation of the complements. Google is in a different position. Its major complements already exist, and it need not worry in the short term about continuing the flow. For content, we have decades of music and movies that can be digitized and then distributed, with advertising attached [and data scraped for profiling]. A wealth of other works awaits digitizing — books, maps, visual arts, and so on. If these run out, Google and other Internet companies have hit on the concept of user-generated content and social networks, in which the users are sold to each other, with yet more advertising attached [and data mining]. So, on the whole, Google can continue to do well even if it leaves providers of its complements gasping like fish on a beach.

To understand how this profound shift affects all artists, I want to turn our attention to the late 1960s, when I was working as a tour manager for The Band — and occasionally for Bob Dylan, when it suited his aloof lifestyle. Whenever a musician complains about how he or she has been screwed by YouTube or Spotify, the standard response is something like this: "Oh, the music business has always screwed the musicians. What else is new?" But this is just not true. The music business worked remarkably well in the 1960s and 1970s. Everyone got paid, you made a decent return on your investment of time and toil as an artist, and record companies really helped you build a career. The individual artist was able to leverage the global distribution networks of the record companies, which in turn were able to nurture artists' careers over multiple albums. Today the business is very different, as the musician David Byrne explained in a *New York Times* op-ed piece: "This should be the greatest time for music in history – more of it is being found, made, distributed and listened to than ever before... Everyone should be celebrating – but many of us who create, perform and record music are not... I myself am doing OK, but my concern is for the artists coming up: How will they make a life in music?"

If we are to have a moral framework for the digital economy, we need to answer this question.

CHAPTER TWO

Levon's Story

Good times don't last long sometimes.
— Levon Helm

1.

I had turned in my final exams at Princeton in early May of 1969 and did not wait around to attend my graduation later that month. My father, a fellow Princetonian who was deeply invested in my matriculation, had died of cancer three years before, at the age of fifty-eight. He was an antitrust lawyer fighting the government on behalf of monopolies such as Dresser Industries, the oil-field services giant, and I never got to ask him if he was happy with his life choices. I was eager to go to Los Angeles because I had gone to work for The Band four months earlier, setting up a recording studio for them in the hills above Sunset Plaza. I didn't have a long view of my career, but in late May, as I followed Robbie Robertson and Levon Helm down to the pool house in Sammy Davis Jr.'s "pad" (complete with a giant bed and mirrors on the ceiling), I felt I was at the right place at the right time.

Robbie and Levon played for me what they had recorded during the three months I had been away finishing my course work. The first tune was "The Night They Drove Old Dixie Down," and at the song's end, tears welled up in my eyes. James Agee and Walker Evans had collaborated on *Let Us Now Praise Famous Men,* and it had opened a window for me into the life of the sharecropper. "Dixie" was like a musical version of that book. Even the black-and-white Elliott Landy portraits of The Band taken for the album hark back to the work of Walker Evans. The song gave me an understanding of Levon's world that would last me the rest of my life. I never viewed the South with the same eyes again after that night.

In the summer of 1965, I had started working part-time as a road manager for Albert Grossman, who was at the time the most important manager in the music business. His clients included Bob Dylan, Peter, Paul & Mary, Paul Butterfield, Odetta, and the Jim Kweskin Jug Band. I carried the Jug Band's guitars, banjos, fiddles, and washtub bass around the fields of the Newport Folk Festival on the epic July weekend when Bob Dylan "went electric" and angered the folk music establishment. For the following six years, I worked for Grossman, first as a weekend roadie, which gave me some money while I completed college, and then as a full-time tour manager for The Band, starting in the spring of 1969. That spring marked their debut at Bill Graham's Winterland Ballroom, in San Francisco.

Levon Helm, The Band's drummer, was born in May of

1940 in the middle of Arkansas cotton country, where his father eked out a living on a small plot of land in a town called Turkey Scratch. He was surrounded by music from the start — his father played the mandolin — but more important, he grew up in a crucial period of music history. As Levon would later explain in a film we made together called *The Last Waltz,* "It's kind of the middle of the country, so if bluegrass or country music — if it comes down to that country and it mixes with blues and it dances, then it's called rock and roll." So Levon came into his adolescence just as a generation of rockabilly musicians five or so years older than he was — led by Elvis Presley, Carl Perkins, and Jerry Lee Lewis — was making the blues accessible to a much wider audience in the still-segregated South. Eventually Levon began performing with his sister as a country act and then got to sit in on drums with Conway Twitty. He told me stories of watching the legendary blues harmonica player Sonny Boy Williamson play in West Memphis, Arkansas, for a fee of $10 per night. Williamson must have known that unless you were really lucky, it was hard to get rich playing music. Levon's first bit of luck came courtesy of a musician from Fayetteville named Ronnie Hawkins, who drafted the seventeen-year-old Levon into his band.

Robbie Robertson, Garth Hudson, Richard Manuel, and Rick Danko eventually joined Levon in Hawkins's band. At some point they broke with Hawkins and went out on their own as Levon and the Hawks. It was then that they crossed paths with my boss, Albert Grossman.

One week after Dylan's 1965 Newport Folk Festival appearance, Mary Martin and Dan Weiner of the Grossman office showed up at Tony Mart's, a roadhouse on the Jersey Shore, and told the Hawks that Bob Dylan wanted them to back him at two concerts: one at Forest Hills Stadium and the other at the Hollywood Bowl. Levon was skeptical and asked who else was going to be on the bill at the Hollywood Bowl; Danny said it was just Bob. The notion that someone with one minor hit on the radio ("Like a Rolling Stone") could fill the seventeen-thousand-seat Bowl strained credulity. When it came to the world of folk music, the Hawks had been living in a parallel universe, and Robbie had to convince Levon that it couldn't hurt for the two of them to just check it out for one gig while the rest of the boys held down the job at the roadhouse. Robbie and Levon went up to New York, rehearsed for a day with Bob (alongside Al Kooper on organ and Harvey Brooks on bass), and played in front of 7,500 fans at Forest Hills Stadium. Dylan's compromise to appease the folkies was to play the first half of the show by himself with his acoustic guitar, then bring the group onstage to close out the night.

My first encounter with Levon's band was later that same year, at their October Carnegie Hall performance. Dylan followed the same formula he'd adopted at Forest Hills, playing the first half acoustic and then coming out with Levon and the Hawks for the rock-and-roll second act. The music was fiery, passionate, outrageous — maybe even dangerous. Unlike the Newport performance (at which Dylan played with an unrehearsed pickup band), the music was tight, and Bob was in top form. He joked between songs and jumped around while Robbie Robertson threw out amazing guitar solos. Rick and Levon provided a rhythm section that anchored the whole group with a driving intensity, bringing songs such as "Maggie's Farm," "Highway 61 Revisited," and "Like a Rolling Stone" to a level Bob had never achieved on his recordings. Maybe it was the august atmosphere of Carnegie Hall, but the audience was respectful of the revolutionary new synthesis of folk's lyricism and rock's power. During the afterparty at Albert Grossman's Gramercy Park town house, the mood was mellow yet congratulatory: Andy Warhol and Edie Sedgwick were just some of those paying court to Dylan and his crew. For Levon and the Hawks, it was an introduction into a new world of art and rock and roll. As Ronnie Hawkins had once promised, they were "farting through silk."

But the fun was not to last. As soon as Bob and the Hawks began to tour the big arenas, pushback from the folkies escalated. The rock set would regularly be met with boos and shouts of "Traitor." For Levon, it must have been dispiriting for other reasons as well: he was no longer the leader of the band, and he hardly ever got to sing backing vocals. After a year or so, he quit and went back to Arkansas while the rest of the group toured the world with a backup drummer, Mickey Jones. Eventually, in the spring of 1966, the scene got old for Bob as well, and, tired of the booing, he ended the tour and retired to his home in Woodstock, New York. Yet the dark mood of the moment still could not be escaped. Dylan, hurtling toward "Gates of Eden" along the dirt roads of Woodstock on a Triumph motorcycle, was tempting fate. It was perhaps unsurprising that he crashed, both literally and figuratively. As he later told *New York Times* music critic Robert Shelton, describing the motorcycle accident that changed his life, "It happened one morning after I had been up for three days." It seemed to me that this was a sort of tap on the shoulder, delivering him back from the abyss. So when Bob began to recover under the loving care of his wife, Sara, he retreated into a world of children (three babies in three years), painting, and the quiet music of an earlier era.

The rest of the Hawks followed Bob to Woodstock and asked Levon to come back as well. When he agreed to do so, they formed a new version of the group, and Albert Grossman promised to get them a recording contract. Robbie Robertson, Rick Danko, and Richard Manuel all began to write songs, drawing upon both the example set by Dylan and the kind of rockabilly music they had grown up with. Everyone in Woodstock called them The Band, and the name stuck. As promised, Albert made a deal with Capitol Records, and a young producer named John Simon was hired to produce their debut album.

2.

The Band's experience is an exemplar of what I mean when I say that the music business made sense in 1967. Capitol advanced the group around $50,000 to cover recording costs. That went toward paying for time in the recording studio and a modest salary for the producer. The group probably spent a month in the studio laying down eleven tracks on a record called *Music from Big Pink*, named for the pink house outside Woodstock that The Band had used as a rehearsal studio. (It was also where the famous album *The Basement Tapes* was recorded.)

Capitol wasn't the only record company that was taking artistic chances. The Grossman office also had a wonderful relationship with Warner Bros. Records, which released Peter, Paul & Mary's albums. Warner, run by Mo Ostin, was a very artist-friendly label that did not operate as you might imagine. It was the opposite of the cliché portrayed on TV shows like *Vinyl* and *Empire.* My friend Ron Goldstein, who was the VP of marketing, explained it to me:

> When I came to Warner in 1969, I was immediately struck by the label's philosophy of signing great artists regardless of whether they had a "hit" on their demos or whether they were necessarily going to sell millions of records. It was all about signing exceptional talent, and in some cases, unique artists. Neither Mo Ostin nor Joe Smith signed Ry Cooder, Van Dyke Parks, or Randy Newman for the specific purpose of using them as a draw for other artists. That process happened organically. But Mo and Joe and all of us at Warner were well aware, after a short period of time, of those artists' value to our company aside from any commercial considerations. These artists became iconic in the eyes of more established artists as well as the media and created Warner's reputation as a home for great music. It should also be said that at the same time, Warner/Reprise was signing major commercially successful artists, which created confidence in the label's ability to market these artists and their music. It was a perfect combination.

So why did the business also work commercially? Like the work of Randy Newman, The Band's *Music from Big Pink* was not a big seller. It was played on alternative FM stations in key cities, including San Francisco, Los Angeles, New York, and Boston, but it was not a Top 40 record. And yet because the recording advance was so modest, The Band began to earn royalty checks within the first year of its release. When I came to work for them full-time in the spring of 1969, the sums on the quarterly checks were increasing as word of mouth spread to other countries and George Harrison and Eric Clapton publicly lauded the record. In the summer of 1969, The Band released its second album, which contained some of their greatest music (including "The Night They Drove Old Dixie Down" and "Up on Cripple Creek"). This led to touring, with three sellout concerts at the Winterland Ballroom. We toured pretty much constantly for the next two years. I won't try to sugarcoat life on the road in 1969: as much fun as it was, it also took its toll, and as the person hired to act as a responsible adult and get the musicians up every morning after a long night of partying, I often tangled with both Levon and Richard, the two members who liked to party the hardest.

Members of The Band were what I call middle-class musicians. They did not make the kind of money that the Rolling Stones or Cream made in the late 1960s and early 1970s, but they made a decent living — enough for Levon to buy himself a house and barn in Woodstock. But here is the irony. A musician in 1969 could rationally conclude that it was just as — if not more — remunerative to be a drummer or a singer as it was to be a songwriter. After all, the total songwriter royalties on an eleven-track LP came to twenty-two cents per copy, whereas the performers' royalties could be as high as $2.50 per copy. And as the fun of the road began to consume the lives of Levon, Rick, and Richard, it was left to Robbie Robertson to continue writing songs for the group. Though he only wrote half the songs on *Big Pink,* he was writing almost all the songs by the time they recorded their third album.

Moreover, groups like The Band had some assurance that, if their music was of lasting quality, they could continue to reap financial rewards long after they'd stopped writing new music. When the CD format was introduced in the early 1980s, their record royalties jumped as old fans bought The Band's classic albums on disc. That royalty stream continued right up until the introduction of Napster in 2000.

And then it ended.

It was horrifying to see The Band members go from a decent royalty income of around $100,000 per year to almost nothing. But the songwriters had made a far better bargain with the digital future, even though no one could have imagined that in 1969. The performing rights organizations (PROs) ASCAP and BMI continued to collect money for songwriters from every venue imaginable. If your song was played in a bar, restaurant, or retail store over the Muzak system, you got paid. If it was used in a commercial, film, or TV show, you got paid. If it was performed live by some other artist, you got paid. And as the number of places using music to relax or excite customers increased, the songwriters' income increased. Unfortunately, there was no organization representing the musicians other than the American Federation of Musicians, which was completely oblivious to the changes digital culture was about to wreak on the business. Even today, performers do not get paid when their songs are played on the radio unless they also wrote those songs.

For fans of The Band, the imbalance in royalty distribution has become a point of contention, anger, and sadness. Levon was diagnosed with throat cancer in 1999, and his ability to sing was profoundly compromised. He wrote a book about his life in which he expressed his anger at Robbie for not sharing his songwriting income. But I was in Woodstock every day in 1969, 1970, and 1971. Robbie Robertson got up every morning and went into his studio and wrote songs until lunchtime; sometimes he would go back after lunch as well. Levon and Richard slept in. For siding with Robbie, I also received Levon's scorn.

Levon's medical bills were high, so he had no choice but to begin putting on a series of concerts in his barn, called the Midnight Rambles, in early 2000. At first he just played drums and got some friends to do the singing. Eventually he was able to do some singing, but it was all pretty painful. Throughout this time, while Levon was barely scraping by, the recordings of The Band continued to be listened to by new generations of musicians, including Mumford & Sons. But because fans listened on pirate sites or on YouTube, Levon had no income stream from this amazing catalog. When he died in 2012, his friends held a benefit at the Izod Center outside New York City so his wife, Sandy, could hold on to the house in Woodstock. Here is the human cost of the Internet revolution.

Now the prominent music blogger Bob Lefsetz has argued that Levon's tragic reality just has to be accepted:

> It's 2015 and not only have recording revenues declined, the whole world of music has gone topsy-turvy. Yes, there are a few superstars who base their careers on successful recordings, but everybody else is now a player, destined to a life on stage. This ain't gonna change, this is the new reality. You can make an album, have fun, but don't expect people to buy it or listen to it. The audience wants an experience. You're better off honing your presentation than getting a good drum sound on hard drive. Your patter is more important than the vocal effects achieved in the studio. You're back to where you once belonged, a performer. Be ready for a life on the road.

Lefsetz is saying that the only way musicians can get paid is the same way they got paid in the seventeenth century: rent a room, lock the doors, and make people pay to get in. By 2020 there will be six billion Internet-enabled smartphones in the world, and how can it be that the arrival of digital networks composed of billions of music fans has not been a boon to musicians?

3.

Even though we were assured by pundits such as *Wired* magazine's Chris Anderson, who wrote his seminal article "The Long Tail" in 2004, that digital abundance would mean a much more democratic distribution of the spoils of the digital age, that notion has turned out to be willful self-deception. The long tail is a myth, a fact evidenced by the current music business, in which 80 percent of the revenue is generated by 1 percent of the content. Even at the height of the early blockbuster era, spawned by Michael Jackson's *Thriller*, 80 percent of the revenue was spread among the top 20 percent of the content. So even in a different winner-takes-all scenario, the revenue was spread out among more artists than it is today. Economists have noted that winners "take all" in many sectors (including hedge funds), and that this has clearly contributed to global income equality, but in the digital media business it seems especially Darwinian. In a world where four hundred hours of video are uploaded to YouTube every minute of every day, the commodification of what was once considered an art (or at least a craft) has become inevitable. For all the stories promoted by Google about YouTube millionaires, the traffic statistics tell another story. Most YouTube videos have fewer than 150 views.

The same thing happens in the streaming music business, where in 2012 ad-supported services such as Spotify paid artists $0.0048 per track. One hundred thousand people listen to your track, and you make less than $500. YouTube is now the world's dominant audio streaming platform, dwarfing Spotify and virtually every other service. Yet it pays artists and record companies less than a dollar per year for every user of recorded music, thanks to rampant piracy on its site. The problem has gotten so bad that, in 2015, vinyl record sales generated more income for music creators than the billions of music streams on YouTube and its ad-supported competitors. In 2015, after years of battling pirates, Prince said in an interview that the Internet "was over for anyone who wants to get paid."

So in the end Levon Helm's problems are our problems. He was able to make great music because he stood on the shoulders of artists like Robert Johnson, Bessie Smith, Hank Williams, Muddy Waters, Maybelle Carter, Buddy Holly, and many others. Some of them died young from tragedy or self-destruction. But most of them got to live long lives making a joyful sound. Sustainable cultures are made by generations of artists able to ply their trade. We value the artists who came before us. Good film directors quote shot sequences from films they admire. And yes, there are small, vibrant communities of art and culture that are still living the life that cultural historian Jacques Barzun described as a renaissance: "The feverish interest, the opposition, and the rivalry among artists working, comparing, and arguing, generate the heat that raises performance beyond the norm." Go to a traditional folk music festival. The quality of the playing and singing will blow your mind. But like the rise in vinyl record production, house shows, and other aspects of hipster culture, it is quintessentially "analog" — the sonic equivalent of the farm-to-table movement. The great electronic musician and producer Brian Eno, who has been working in funky analog studios in West Africa, has begun to question the very raison d'être of digital recording, which, thanks to Auto-Tune (the tech tool that allows engineers to correct singers with bad pitch), makes it possible to turn a second-rate singer into a diva: "We can quantize everything now; we can quantize audio so the beat is absolutely perfect. We can sort of do and undo everything. And of course, most of the records we like, all of us, as listeners, are records where people didn't do everything to fix them up and make them perfect." Tech's perfection tools do not make for human art.

The fiddlers and banjo pickers at the Union Grove Old Time Fiddlers Convention, in Union Grove, North Carolina, preserve their music as a hobby. Most of them have "day jobs" and play on weekends. But that doesn't mean today's professional artists develop their music in a vacuum. Clearly a radical artist such as Kendrick Lamar has spent a lot of time listening to artists who have gone before him – in Lamar's case, they include the free jazz musicians Sun Ra and Charles Lloyd – yet most of his listeners have no idea who these previous artists are. The way Lamar might like to pass great music to the next generation is in fundamental conflict with the tools available to the current artist. The ideal system imagined by the long tail has not come into being, even though the great majority of history's recorded music is available online. What the search engine does is reward only the newest and most popular content. It does not push you down the long tail.

I recall an evening at Levon Helm's house in Woodstock in 1969 when he was trying to teach a young kid (me) about the concept of slowness in music. In many ways rock and roll seemed to be about speed—Little Richard's rapid piano chords, Chuck Berry's "Maybelline"—yet Levon was enamored with the idea of just how slowly you can play a song and still keep the rhythm moving forward. He put the record Ray Charles Live on the turntable and played "Drown in My Own Tears." The beat of the song is unbelievably slow. The drums and bass seem to be in a time warp. Each line plays itself out as if it's resisting coming to the end of the bar. "Drown in My Own Tears" is, of course, a blues tune, but the slowness exaggerates the sadness to such a level that you wonder whether Ray can get through the song. It reminds me of the note Dmitry Shostakovich wrote to the players of his last string quartet (no. 15), completed in 1974. Each movement is marked adagio, and he wrote, "Play the first movement so that flies drop dead in mid-air and the audience leaves the hall out of sheer boredom." That's what the live recording of "Drown in My Own Tears" feels like, except that when Ray finally finishes, his audience goes absolutely crazy with joy.

Levon gave me a gift that night. But how do we keep this sense of the arc of history and culture when we are flooded with the fierce urgency of now? How do we really take advantage of the Internet's original purpose to decentralize its control and deepen our knowledge base?

Perhaps the answer lies in understanding the countercultural roots of the Internet.

CHAPTER THREE

Tech's Counterculture Roots

He was dealing lightning with both hands.
— Computer scientist Alan Kay

1.

It was raining like hell outside, and Doug Engelbart was pacing nervously on the stage. A tall, fit forty-three-year-old wearing a crisp white shirt and blue tie, with streaks of gray showing in his neatly parted hair, he looked like he could work for NASA or the Defense Department. And he did, in the sense that for the past several years the Stanford Research Institute (SRI), in Menlo Park, California, had funded his quixotic quest to invent the future.

In three hours the auditorium would be filled with the best computer scientists in the world, all gathered for the annual conference of the 1968 Association for Computing Machinery and the Institute of Electrical and Electronics Engineers (ACM/IEEE). Computing machinery! As far as the association was concerned, humanity had not left the industrial age, yet their members were about to enter the information age. Doug called the system he had built the oN-Line System (NLS), and in the 100-minute demonstration that would follow he planned to introduce the world to (in the words of Engelbart's biographer Thierry Bardini) "windows, hypertext, graphics, efficient navigation, command input, videoconferencing, the computer mouse, word processing, dynamic file linking, revision control, and a collaborative real-time editor." But for the moment no one was sure that what would later be called the Mother of All Demos would work. Doug had told someone at NASA earlier in the week that he was going to show the system publicly – "Maybe it's a better idea you don't tell us, just in case it crashes," the NASA employee advised him. Doug's chief engineer, Bill English, had been a theatrical stage manager and knew that the demonstration had to be ready as soon as the audience showed up. But what a show. Let Doug explain.

> We set up a new Swiss video projector aimed at a 22-foot screen. On the right side of the stage I sat at a Herman Miller console with a display, keyboard, mouse and special keyset input device. We built special electronics that picked up the control inputs from my mouse, keyset and keyboard and piped them down to SRI over a telephone hookup. We leased two microwaves up from our laboratory in SRI, roughly 30 miles. It took two additional antennas on the roof at SRI, four more on a truck on Sky-

> line Boulevard and two on the roof of the conference center. On the console on the stage there was a camera mounted that caught my face. Another camera mounted overhead, looked down on the workstation controls. In the back of the room, Bill English controlled use of these two video signals as well as two video signals coming up from SRI that could bring either camera or computer video.

When the show started it was as if Engelbart had arrived from the future, "dealing lightning with both hands." The effect on the thousand people gathered for the conference was revolutionary. Imagine the first performance of Stravinsky's *The Rite of Spring* but without the boos and walkouts. People were thunderstruck by this radical upending of what a computer could be. No longer a giant calculation machine, it was a personal tool of communication and information retrieval.

2.

It is not an exaggeration to say that the work of Steve Jobs, Bill Gates, Larry Page, and Mark Zuckerberg stands on the shoulders of Doug Engelbart. Yet Engelbart's vision of the computing future was different from today's reality. In the run-up to the demonstration, Bill English had enlisted the help of *Whole Earth Catalog* publisher Stewart Brand, who had produced the Acid Tests with Ken

Kesey two years earlier. Engelbart felt that Brand might help make his show into a multimedia event. Kesey and Brand's LSD festival had forever cemented San Francisco's link to what Fred Turner in his book *From Counterculture to Cyberculture: Stewart Brand, the Whole Earth Network, and the Rise of Digital Utopianism* describes as the New Communalists. Engelbart himself had taken acid twice under the supervision of a Stanford psychology PhD, Jim Fadiman, at the International Foundation for Advanced Study, the Bay Area research hub for academic LSD studies, which were legal until 1967. Geeks on acid, dreaming of the future. But financed by the military-industrial complex. Complicated.

The Whole Earth Catalog, subtitled *Access to Tools,* begins: "We are as gods, and we might as well get used to it." Pretty bold mission statement. Was this a church? No, but the sense was that the New Communalist needed tools to create an individual identity free from the hidebound institutions of contemporary society, which was not an easy task. The introduction to the catalog continues, "A realm of intimate personal power is developing – power of the individual to conduct his own education, find his own inspiration, shape his own environment and share his adventure with whoever is interested." Engelbart got this vision. NLS would provide a tool for individual empowerment: the user could access the world's knowledge, create inspiring content, and share it with anyone. Less than two minutes into the San Francisco demo, Engelbart said, "If in your office you as an intellectual worker were supplied with a computer display backed up by a computer that was alive for you all day and was instantly responsive to every action you have, how much value could you derive from that?" Engelbart had built a working prototype of what we today would easily recognize as a contemporary Internet device — fifteen years before the introduction of the Apple Macintosh.

The next year Engelbart took a team from the Stanford Research Institute to the Lama Foundation commune, north of Taos, New Mexico. It was Stewart Brand who suggested that Lama might provide an atmosphere, as John Markoff wrote, "to create a meeting of the minds between the NLS researchers and the counterculture community animated by the *Whole Earth Catalog*." The land outside Taos was full of alternative communities — Morningstar East, Reality Construction Company, the Hog Farm, New Buffalo, and the Family, to name a few. Steve Durkee and Steve Baer, both disciples of Buckminster Fuller and close friends of Stewart Brand, ran Lama, and the architecture of the buildings hewed closely to Bucky Fuller's geodesic dome design.

Fuller believed that what society needed was not more specialization but a new type of generalist, whom he called a comprehensive designer. For Bucky the problem of humanity's survival was one of design, and he thought the "artist-scientist" could solve it:

> If man is to continue as a successful pattern-complex function in universal evolution, it will be because the next decades will have witnessed the artist-scientist's seizure of the prime design responsibility and his successful

> conversion of tool-augmented man from killingry to advanced livingry — adequate for all humanity.

The irony, of course, was that Brand understood that most of the financing for Engelbart's work was flowing from the fount of "killingry" — the government's Defense Advanced Research Projects Agency, or DARPA. DARPA was a direct response to the Soviet launch of *Sputnik 1* in 1957 and was set up to fund technology research projects that would expand the frontiers of expertise beyond the immediate and specific requirements of the military and its laboratories. It was an extremely flat organization, characterized as 100 scientists and a travel agent, that set out to give major university computer science labs the economic support to conduct basic research that would lead to US technological superiority in computers and networked connectivity. One of its first successful projects was the progenitor of the Internet — ARPANET, the world's first packet switching network developed in 1962 between four university campuses.

Here is the paradox that libertarians just don't get: the Internet was conceived and paid for by the US government. It was not a product of the free market as we think of it today — the realization of some young entrepreneur's dreams. It was painstakingly researched and executed by a bunch of academics for whom IPO billions weren't a reason to work. Rather, these people were fundamentally convinced that they could make the world a better place with their inventions. Every piece of code — HTML,

TCP/IP – was donated to the ARPANET project royalty-free. Of course DARPA had its own reasons for funding Doug Engelbart's research, deeply interwoven with Cold War paranoia and post–nuclear attack "survivability," but that was irrelevant to the purpose and the idealism of Engelbart, Brand, Vint Cerf, Tim Berners-Lee, and a host of other geniuses who made the Internet. But ultimately the connection with the military led to the undoing of Engelbart's NLS vision.

By 1969 the antiwar demonstrations outside the SRI building were a daily occurrence. Inside, the research team, which was growing quickly – thanks to increasing DARPA investment after the successful San Francisco demo – began to break into two factions: computer geeks and countercultural humanists. Engelbart had a hard time holding these groups together. The hippies would go off on retreats and not invite the committed DARPA scientists. In 1970 Xerox hired one of the leaders of DARPA, Bob Taylor, to form a new lab called the Palo Alto Research Center (PARC). Taylor's first move was to poach Bill English, the chief NLS engineer. Within months they had built a team, among whom was computer scientist Alan Kay, that would create a commercial version of Engelbart's vision.

What is so important about Engelbart's legacy is that he saw the computer as primarily a tool to augment – not replace – human capability. In our current era, by contrast, much of the financing flowing out of Silicon Valley is aimed at building machines that can replace humans. In a famous encounter in 1953 at MIT, Marvin Minsky, the father of research on artificial intelligence, declared: "We're going to make machines intelligent. We are going to make them conscious!" To which Doug Engelbart replied: "You're going to do all that for the machines? What are you going to do for the people?"

3.

Engelbart's vision of the world was about to be eclipsed by a more commercially oriented one. Stewart Brand began hanging out with the PARC team and eventually wrote an article for *Rolling Stone* called "Spacewar: Fanatic Life and Symbolic Death Among the Computer Bums." It starts like this.

> Ready or not, computers are coming to the people.
>
> That's good news, maybe the best since psychedelics. It's way off the track of the "Computers — Threat or menace?" school of liberal criticism but surprisingly in line with the romantic fantasies of the forefathers of the science such as Norbert Wiener, Warren McCulloch, J.C.R. Licklider, John von Neumann and Vannevar Bush. The trend owes its health to an odd array of influences: The youthful fervor and firm dis-Establishmentarianism of the freaks who design computer science; an astonishingly enlightened research program from the very top of the Defense Department; an unexpected market-Banking movement by the manufacturers of small calculating machines; and an irrepressible midnight phenomenon known as Spacewar.

Brand here was tying the networked computer revolution directly to the counterculture he was championing in the *Whole Earth Catalog*. And while he proclaimed that this revolution was the best news "since psychedelics," he was fully aware that PARC was filled with video-gaming freaks financed by the Pentagon. The image makers at Xerox headquarters back East practically had a heart attack and decreed that there should be no more reporters at PARC. But Alan Kay – the young team leader who had envisioned the Dynabook, the first iteration of a PC, while he was still a PhD candidate – didn't care. They were proud to fly their freak flag at PARC. Kay's real passion was to design an educational tool using Engelbart's basic ideas but putting an enhanced emphasis on the graphical user interface (GUI). Despite resistance from some of PARC's management, Kay managed to assemble a small team to build the Alto – the first *real* personal computer. It had a mouse, a GUI that looks much like the one on your Mac, an Ethernet connection so it could link to other computers, and a printer. Xerox's first reaction was "Couldn't it be used by four people?" The company's executives missed the whole point: it was a personal communication and creativity device. The first demonstration had *Sesame Street*'s Cookie Monster holding a cookie with one hand and then bringing up the letter C in the other. Kay knew that a first grader could manipulate a computer that had a simple GUI and a mouse.

John Seely Brown, who was to become director of PARC, told me that two principles were at the center of the PARC philosophy. To begin with, Bob Taylor had brought the core directive of ARPANET with him to PARC. As Brown notes, "Decentralization was fundamental to ARPANET in the sense that a nuclear strike on a single city could not bring down the entire network." Everything done at PARC, from the Alto to Bob Metcalf's Ethernet architecture, was geared to making a decentralized network of personal computers function efficiently. This was new. The second core principle flowed from Alan Kay's Dynabook. As Brown says, "The Dynabook and then the Alto were inspirations meant to empower the artistic individual." When Brown first started working with Kay, he was playing music on the Alto and working with Stanford's Center for Computer Research in Music and Acoustics. These two innovations – a decentralized network and a personal creativity machine – became the core of the Internet revolution. Despite the revolution's value, the culture of PARC – which Brown describes as being one of "long hair and no shoes" – was a bit of a mismatch for Xerox. But the culture clash was not just about hippies versus suits. Brown explains: "Xerox built big complicated stuff that sold for $250,000 a unit and came with three-year guarantees. What was the chance that any of the PARC stuff could ever be sold through the Xerox channels? Zero." So the decision was made to try to partner with Apple.

Almost every version of the story of Steve Jobs visiting PARC for a demonstration in December of 1979 is wrong. It is usually said to epitomize the complete failure on Xerox's part to understand what they had invented. A bit of background: Apple had successfully launched the Apple II computer in April of 1977. It was an instant hit, and between September of 1977 and September of 1980, yearly sales grew from $775,000 to $118 million, an average annual growth rate of 533 percent. But Jobs refused to sit still. He had heard about the Xerox Alto and had worked out a deal with Xerox in which he would sell them perhaps as much as 5 percent of Apple in return for a licensing agreement to all the PARC technology. But in a true sign of how dysfunctional Xerox management was, it did not inform the PARC executives of the pending stock transaction. Xerox merely informed them that Jobs was going to come to PARC and that they should give him a demonstration.

Alan Kay and the group complied with the request, but they did not "open the kimono" in any way. Jobs left in a rage and called Xerox. The next day he returned, and a chastened PARC team showed him everything. Jobs went back to Cupertino and called a board meeting, saying he had to build a new computer based on the PARC architecture and that it should not be backward-compatible with the existing Apple II. The board thought he was crazy, but Jobs applied his charisma – his "reality distortion field" – and got his way. Xerox got its Apple shares, and in December of 1980, Apple went public at $22 per share. Xerox's holdings were instantly worth millions.

The first version of a computer using the PARC architecture, the Lisa, was a commercial failure, but when Jobs introduced the Macintosh in an iconic advertisement that aired during the 1984 Super Bowl, the long-awaited vision of the future arrived. The tragedy for Xerox was that two years later, the Xerox CFO sold all its Apple stock. Imagine what it would have meant to the company if it had held on to 5 percent of Apple, which would now be worth about $32 billion. In 1985, after the debut of the Macintosh, Microsoft quickly introduced Windows, an operating system that totally mimicked the Macintosh. Whatever advantage Apple had was quickly extinguished, and Steve Jobs was forced out of the company.

Jobs immediately set out for revenge on his old company by building a new computer called NeXT. Not long after that, a twenty-nine-year-old English engineer, Tim Berners-Lee, took up a position at the Conseil Européen pour la Recherche Nucléaire (CERN). The Internet at this point was purely an academic research network linking physicists around the world and allowing them to share research documents, and CERN was the largest European node of the network. Finding documents was getting increasingly dicey as the network got more popular, so Berners-Lee began to work on the concept of hypertext as a way for researchers to link directly to other documents in their references. Coincidentally, he and Steve Jobs crossed paths at the very moment of the birth of the World Wide Web.

In late 1988, Mike Sendall, Berners-Lee's boss at CERN, bought a NeXT cube to evaluate. A few months later, Berners-Lee had an idea to combine hypertext and the Internet and submitted it to Sendall. Sendall approved the proposal and gave Berners-Lee the NeXT cube to play with. Berners-Lee's colleague Robert Cailliau told the story:

> Tim's prototype implementation on NeXTStep is made in the space of a few months, thanks to the qualities of the NeXTStep software development system. This prototype offers WYSIWYG browsing/authoring! Current Web browsers used in "surfing the Internet" are mere passive windows, depriving the user of the possibility to contribute. During some sessions in the CERN cafeteria, Tim and I try to find a catching name for the system. I was determined that the name should not yet again be taken from Greek mythology... Tim proposes "World-Wide Web". I like this very much, except that it is difficult to pronounce in French.

So the World Wide Web changed everything, and Steve Jobs's "what you see is what you get" (WYSIWYG) interface helped make it possible. Hyperlinking and open access became easy. But for Berners-Lee today there is some regret as he looks back at the birth of the Web. The Web was built to decentralize power and create open access, yet, he noted, "popular and successful services (search, social networking, email) have achieved near-monopoly status. Although industry leaders often spur positive change, we must remain wary of concentrations of power." Notice the tentative voice: he doesn't mention Google and Facebook by name. Tim Berners-Lee never got rich on his invention. He gave it to the world for free, so he remains dependent on research funding from giant corporations.

4.

At about the same time that Berners-Lee was creating the World Wide Web, Stewart Brand was creating the Whole Earth 'Lectronic Link (WELL). When it started, in 1985, it consisted of one computer in the Whole Earth office and a large group of Bay Area computer enthusiasts who could log on through a dial-up modem and have real-time text conversations on a variety of subjects. In his 1972 *Rolling Stone* article, Brand identified these enthusiasts as "hackers," saying, "The hackers are the technicians of this science — It's a term of derision and also the ultimate compliment. They are the ones who translate human demands into code that the machines can understand and act on. They are legion. Fanatics with a potent new toy." The early WELL guidelines were very clear about intellectual property, as stated on the log-on screen: "You own your own words. This means that you are responsible for the words you post on the WELL, and that reproduction of those words without your permission in any medium outside the WELL's conferencing system may be challenged by you, the author."

But in 1989 something weird happened. The notion of the "hacker ethic" became a contested trope. It started with an online forum on the WELL organized by *Harper's Magazine*. The subject was hacking, and Paul Tough, a *Harper's* editor, had recruited Brand and a few of his most important WELL members, including Howard Rheingold, Kevin Kelly, and John Perry Barlow, to participate. Barlow, a shaggy, bearded man with a fondness for colorful cowboy shirts, is a true American character. He is a self-confessed failed Wyoming rancher, a former Catholic mystic, and a former Grateful Dead lyricist. He was also an early proponent of cyberspace as the new American frontier – as lawless as Wyatt Earp's Tombstone. Barlow wanted it to stay that way, but his romantic notion of the hacker as countercultural brigand was about to be confronted with something more real and more dangerous.

In the spirit of a hacker forum, the Harper's editor invited two real hackers, identified only as Acid Phreak and Phiber Optik, to join the discussion. The debate over the definition of hacker soon got pretty heated. Fred Turner set the scene in *From Counterculture to Cyberculture*:

> Among WELL regulars like... Barlow, hackers were cybernetic counterculturalists, creatures devoted to establishing a new, more open culture by any electronic means necessary. For Acid Phreak, hackers were break-in artists devoted to exploring and exploiting weaknesses in closed and especially corporate systems.

Barlow kept insisting that a computer network was like a small Wyoming town where people left their doors unlocked, but Acid Phreak would have none of it. In a fit of youthful fury, he hacked Barlow's credit report and posted it on the WELL. Now the young hackers had Barlow hooked. He wanted to hang out with them. Maybe all his Jesse James fantasies were tied up in these kids, but it ended badly. On January 24, 1990, the Secret Service raided the apartment where Acid Phreak was living with his mother. By the end of the day Acid Phreak, Phiber Optik, and a third hacker, Scorpion, were in a New York City jail, accused of hacking the main AT&T computer system.

And that is where Brand, Barlow, and their fellow communards changed course. Instead of recalibrating the hacker ethic in line with their earlier goals, they embraced the criminals on the theory that theirs was really a victimless crime. Tell that to AT&T, which had to spend millions restoring its system. Barlow went on to form the Electronic Frontier Foundation, which has never met a hacker it couldn't defend. No more defense of intellectual property for these boys – "Information wants to be free." But of course that is a lie. As the science fiction writer Cory Doctorow has pointed out, "The desires of information are totally irrelevant to the destiny of the Internet, the creative industries, or equitable society. Information is an abstraction, and it doesn't 'want' anything." Even the most radical of the Valley's libertarians, Peter Thiel, didn't believe the "information wants to be free" nonsense, noting, "Every great business is built around a secret that's hidden from the outside." The EFF is a symbol of our current era, a time when hacker ideology trumps common sense. The group has gone as far as to defend "revenge porn" website owners who were ordered by courts to take down their sites.

In the end Stewart Brand abandoned his communard dreams for a new calling: corporate consultant. He had gotten a taste of the power of a social network at the WELL. If a company could sponsor an online community and if it could convince its customers that they were engaging in social rather than economic activity, then they could increase customer allegiance and their own profits. From this insight flowed the Global Business Network. Forget going back to the land – there was gold in preaching that Whole Earth message in the suites of the Fortune 500. The corporate conquest of the Web had started.

CHAPTER FOUR

The Libertarian Counterinsurgency

We are in a deadly race between politics and technology.
— Peter Thiel

1.

George Gilder was down on his luck. Sweating like a pig in a humid office with a broken air conditioner, he was working in 1972 for Ben Toledano, a failed candidate for mayor of New Orleans who believed he could become Louisiana's next senator. Gilder — thirty-three, a graduate of Exeter and Harvard, and a speechwriter for Nelson Rockefeller and Richard Nixon — was toiling for beer money with a second-rate candidate who had no more chance of becoming senator than he had of becoming mayor. After all, Toledano had just joined the Republican Party three years before, right after he gave up his membership in the Louisiana States Rights Party, which advocated extreme right segregationist policies. Since Toledano would only pay

Gilder for four hours a day, the younger man had his afternoons and evenings free to think about what a damn mess his life had become. In the midst of this stew of anger and self-pity he came to the conclusion that his plight was all the fault of the women's movement. So he set out to write a book called *Sexual Suicide,* which would wake the country up to the poison in its midst. A review of the book in *Kirkus Reviews* states his theme:

> Women's Lib and its goals — abortion on demand, child-care centers, equal pay for equal work — will be the ruination of us all. Anything that takes the woman out of the home will add to the male sense of redundancy, impotence and rootlessness; take away his age-old role as protector and provider and he will turn to drugs, pornography, marauding, rape and killing.

To Gilder it was simple. Welfare and feminism had turned men into a subservient race, no longer the hunter-gatherer but the chump. As Gilder was to discover, making outrageous claims was just what he needed to get himself out of his obscure hole in New Orleans. The National Organization for Women named him Male Chauvinist Pig of the Year, and William F. Buckley invited him to appear on *Firing Line.* He decided he would make himself "America's number one anti-feminist."

But just being an anti-feminist was not enough for Gilder's ambition, and he began to write for the editorial page of the *Wall Street Journal* about the supply-side economics theories that Jude Wanniski and Arthur Laffer were putting forward as a conservative response to Keynesian demand-side economics, which had become conventional wisdom. Wanniski and Laffer believed that high tax rates and government regulation were the greatest impediments to growth, impeding capital formation and investment. They argued that if you lower taxes and cut regulations, the rich will invest more, thereby producing more goods at lower prices and stimulating demand. All this got packaged in a book Gilder wrote called *Wealth and Poverty,* which came out in 1981, just as the Reagan administration took office.

The economy had been struggling through seven years of stagflation — a toxic mix of high inflation and stagnant growth. The notion that big government was the problem became the basis of Reagan's campaign. Gilder's book appealed to Reagan's sense that what really ailed America was the fault of "welfare queens." In a 1994 interview Gilder said, "The so-called 'poor' are ruined by the overflow of American prosperity. What they need is Christian teaching from the churches. . . . We have no poverty problem strictly speaking, we have a desperate problem of family breakdown and moral decay." Gilder began to embrace a new set of norms that were championed by Laffer and others but were really driven by the ideas of Ayn Rand — the concept of "the makers and the takers" writ large. They called themselves libertarians and began to imagine an economy free from government regulation and taxes, powered by new technologies such as fiber optics and personal computing.

2.

The sixteen-year-old chess prodigy Peter Thiel, bored by his high school in Foster City, California, was seized with libertarian fervor. The *New Yorker*'s George Packer sets the stage.

> He became a math prodigy and a nationally ranked chess player; his chess kit was decorated with a sticker carrying the motto "Born to Win." (On the rare occasions when he lost in college, he swept the pieces off the board; he would say, "Show me a good loser and I'll show you a loser.") As a teenager, his favorite book was *The Lord of the Rings,* which he read again and again. Later came Solzhenitsyn and Rand.

Like Gilder, Thiel was outraged by feminism. By the time he showed up as an undergraduate at Stanford, his ideas on the subject were very close to Gilder's. Years later, Gilder returned the compliment, as *Forbes* columnist Ralph Benko noted: "When George Gilder, arguably the smartest man in the world, says, as he said to me over dinner recently in Washington, DC, that Peter Thiel is the smartest man in the world . . . pay attention." In a speech, Thiel later explained why he formed the *Stanford Review*: to fight feminism and political correctness on campus. He later made an outrageous statement on the Cato Institute website: "Since 1920, the vast increase in welfare beneficiaries and the extension of the franchise to women — two constituencies that are notoriously tough for libertarians — have rendered the notion of 'capitalist democracy' into an oxymoron."

Staffed by Thiel's fellow libertarians, such as David Sacks, the *Review* encouraged its readers to test the limits of free speech. One law student friend of Thiel's named Keith Rabois decided to take the idea to the limit by standing outside the dorm residence of one of his instructors and shouting over and over, "Faggot! Faggot! Hope you die of AIDS." The university community recoiled, and Rabois eventually withdrew from the law school. Thiel and his coauthor, Sacks, in their book *The Diversity Myth,* came to Rabois's defense, writing, "His demonstration directly challenged one of the most fundamental taboos: To suggest a correlation between homosexual acts and AIDS implies that one of the multiculturalists' favorite lifestyles is more prone to contracting the disease and that not all lifestyles are equally desirable." Published in 1995, ten years before Thiel publicly declared his own homosexuality, the book is an extended rant against gay culture and racial and gender diversity — the "multiculture," to use Thiel's term — on the Stanford campus. He notes, "The multiculture exists to destroy Western culture, and this destruction has been ferocious and indiscriminate." This deep contradiction — a gay man railing against diversity — is at the heart of our understanding of Thiel and the revolution he was about to lead.

Thiel's philosophical hero was Ayn Rand, whose first major novel, *The Fountainhead,* tells a tale of an individualistic architect, Howard Roark, who is depicted as a superior man struggling against the suffocating mob. Begun in 1937, a point when Rand was identifying herself with a small band of anti-Roosevelt partisans such as H. L. Mencken and

Albert Jay Nock, who were the first to use the term *libertarians, The Fountainhead,* along with Rand's later novel, *Atlas Shrugged,* became the urtexts of the libertarian movement. The largest organized anti-Roosevelt group called itself the American Liberty League, which Rand's biographer Jennifer Burns describes as "a secretive cabal of wealthy businessmen hoping to wrest control of the government from the masses." Like her hero Roark, Rand believed the individual should "exist for his own sake, neither sacrificing himself to others nor sacrificing others to himself." Ultimately Rand had an almost Nietzschean will-to-power philosophy, which she handed down to followers such as Thiel. It plays out perfectly in these lines from *The Fountainhead*:

> "Do you mean to tell me that you're thinking seriously of building that way, when and if you are an architect?"
>
> "Yes."
>
> "My dear fellow, who will let you?"
>
> "That's not the point. The point is, who will stop me?"

"Who will stop me." This became the central tenet of Internet disrupters, from Thiel's PayPal right up to Travis Kalanick's Uber.

3.

The ideas of Ayn Rand have not only inspired men like Thiel, they have also found a home in the highest political offices in the country. Paul Ryan, the Speaker of the House, told the *Weekly Standard,* "I give out *Atlas Shrugged* as Christmas presents, and I make all my interns read it." In a video series, Ryan said, "Ayn Rand, more than anyone else, did a fantastic job of explaining the morality of capitalism, the morality of individualism, and this, to me, is what matters most." But as Paul Krugman has written, the Republican elite is "too committed to an Ayn Rand story line about heroic job creators versus moochers to admit either that trickle-down economics can fail to deliver good jobs, or that sometimes government aid is a crucial lifeline."

The notion of altruism and cooperation is not something that Peter Thiel believes in. He is as confident as an Ayn Rand hero that "achievement of your happiness is the only moral purpose of your life." He is not just a good businessman; it's worth noting that at Stanford he earned his degree in philosophy, not in business. From the start, PayPal was pitched as a libertarian philosophy project, an effort to disrupt the credit card–banking system by offering an alternative way to make online payments. PayPal was also committed from the outset to avoiding government intrusion: Thiel does not believe in regulation, paying taxes, or guarding copyright. In addition, PayPal spawned a number of start-ups run by the young men of the PayPal Mafia, who founded YouTube, LinkedIn, Yelp, Palantir, and many other companies.

It's important to understand that while libertarian philosophy may have been regarded as a fringe movement in American politics, epitomized by Ron Paul followers, it has become the mainstream economic philosophy for both

Silicon Valley and the Republican Party, thanks to the Koch brothers. The libertarian belief that the supremacy of the free market is the natural order of things is in reality nothing more than an "imagined order," which the historian Yuval Noah Harari defines in his book *Sapiens* as the shared myths we use to induce cooperation. "In order to safeguard an imagined order," Harari writes, "continuous and strenuous efforts are imperative." Adam Smith's "invisible hand" is no more a law of nature or physics than Moses's Ten Commandments.

Of course Thiel's comment about capitalism and democracy being incompatible points to an even darker part of the libertarian cult that Charles Koch flirted with, sometimes known as anarcho-capitalism or paleo-libertarianism. Its two main philosophers are Murray Rothbard and Hans-Hermann Hoppe. Hoppe wrote a book called *Democracy: The God That Failed,* which makes the argument that we need to return to a more authoritarian form of government. Here is Hoppe's thesis:

> In the United States, less than a century of full-blown democracy has resulted in steadily increasing moral degeneration, family and social disintegration, and cultural decay in the form of continually rising rates of divorce, illegitimacy, abortion and crime. As a result of an ever-expanding list of non-discrimination — "affirmative action" — laws and non-discriminatory, multicultural, egalitarian immigration policies, every nook and cranny of American society is affected by government management and forced integration; accordingly, social strife and racial ethnic

> and moral-cultural tension and hostility have increased dramatically.

That a country in such peril would need a more authoritarian government that could counter the forces of "egalitarian immigration" and "forced integration" feels at once like a throwback to the Jim Crow 1950s and a contemporary statement from someone like Donald Trump. This is in essence the truly elitist theory behind Peter Thiel's thinking. Though his undergraduate degree was in philosophy, Thiel gravitated toward technology and politics. What he believed was that politics was impeding progress and that he needed to find a way to make money without its interference. He wrote in his manifesto on the Cato Institute's website:

> In our time, the great task for libertarians is to find an escape from politics in all its forms – from the totalitarian and fundamentalist catastrophes to the unthinking demos that guides so-called "social democracy".... We are in a deadly race between politics and technology.... The fate of our world may depend on the effort of a single person who builds or propagates the machinery of freedom that makes the world safe for capitalism.

Clearly he saw himself as the person to lead this crusade. He would create a business sector free from taxes, regulation, and copyright that would take the foundation laid down by Doug Engelbart, Stewart Brand, and Tim Berners-Lee in a whole new direction.

4.

The first "machinery of freedom" Thiel, Max Levchin, and Luke Noseck designed was PayPal. In his book *Zero to One: Notes on Startups, or How to Build the Future,* Thiel notes proudly, "Of the six people who started PayPal, four had built bombs in high school." Originally called Confinity, the company was started in December of 1998 as a payment system for the Palm Pilot handheld computer. The ambitious philosophy of PayPal is clearly stated in early remarks from Thiel: "It will be nearly impossible for corrupt governments to steal wealth from their people through their old means," he once told his staff, "because if they try the people will switch to dollars or Pounds or Yen, in effect dumping the worthless local currency for something more secure."

So from the basic Randian concept that governments cannot be trusted with currency, he moved on to set up the basic principles that would constitute the foundation for most new Internet fortunes.

1. Build a proprietary technology that has significant benefits over the competition. Thiel says, "It's always a red flag when entrepreneurs talk about getting 1% of a $100 billion market." He wanted to invest in monopolies, not competitive businesses.
2. Build businesses that have "network effects." Thiel's first two major investments, PayPal and Facebook, both benefit from having millions of users who want to connect with each other. When PayPal was just a

payment system for Palm Pilot, it was a failure. As soon as it became the standard payment system for eBay, it got the network-effect wind at its back.

3. Economies of scale are critical. Google is pretty much unassailable in search-engine advertising because it has huge economies of scale. This leads to the conclusion that there will be very few winners in each sector of tech. The combination of scale and network effects makes it very hard to dislodge the winners, especially if you are in a business like tech, which is so lightly regulated.
4. Branding becomes critical. The brand becomes a promise of value to consumers. Apple gets superior margins because of its brand promise for quality and elegant design. The brand promise also helps you defend yourself against government intrusion. Google's original "Don't be evil" brand promise gave them a patina of social entrepreneurship that helps protect them from accusations of monopoly power tactics.

As John Seely Brown has pointed out, the end of the decentralized Web that Engelbart and the PARC visionaries had imagined occurs at this point, when "we moved from products to platforms, which let the network effect play out in a hub and spoke model." From this point on the economies of scale enjoyed by a platform whose users are measured in the billions becomes the ultimate metric for success. Thiel understood this, and from PayPal, the original founding group began to spread through Silicon Valley after eBay's $1 billion acquisition of the company. Even companies not funded by Thiel adopted his unique view of capitalism. The foundations of an Internet economy that operates with little regulation have been the basis of their fortunes.

5.

But perhaps the greatest beneficiary of the no-taxes, no-regulation regime of the Internet has been Jeff Bezos, the founder and CEO of Amazon. Bezos was schooled in the libertarian ethos by his family. His stepfather, Miguel Bezos, had emigrated from Cuba after the rise of Castro and was an engineer for Exxon in Houston, Texas. Mike Bezos (as he is now known) contributed to Gary Johnson's Libertarian Party presidential campaign in 2012. Jeff Bezos outlined his core libertarian philosophy in an interview for the Academy of Achievement:

> I think people should carefully reread the first part of the Declaration of Independence. Because I think sometimes we as a society start to get confused and think that we have a right to happiness, but if you read the Declaration of Independence, it talks about "life, liberty and the pursuit of happiness." Nobody has a right to happiness. You should have a right to pursue it, and I think the core of that is liberty.

Bezos was working at the hedge fund D. E. Shaw in May of 1992 when the Supreme Court ruled in *Quill Corp. v.*

North Dakota that "the lack of a physical presence in a state is sufficient grounds to exempt a corporation from having to pay sales and use taxes to a state." For Bezos, who studied the growth of the Internet while at Shaw, a light went on. He began to imagine an online retailer that could totally disrupt the local bookstore business. His principles were very similar to Thiel's four guidelines. First he would build a proprietary online platform (even going so far as to patent "one-click" ordering). Then he would harness the network effect, using user recommendations to build individual taste profiles of every customer. He would also create economies of scale by buying in bulk from publishers and negotiating the lowest prices, which individual bookstores could not match. In this effort he could use the wholesaler Ingram in order to list those titles he was not buying from publishers, thereby dramatically expanding the number of books he could offer on Amazon and still avoid charging sales tax. And finally he would build a brand around the name Amazon, which he chose both for its connotation of scale – the world's largest river – and its position in the alphabet: because it starts with *A*, it would be at the top of most lists. By 1994, Bezos had quit his hedge fund job and set Amazon up in Seattle, reasoning that the state was sparsely populated, so the majority of his customers would come from outside the state and therefore would not pay sales tax.

By 1997 Amazon was growing so fast that it was putting many local bookstores out of business. Legislators in several states began to lobby for imposing sales tax on Amazon. The economist Dean Baker has estimated that

Amazon's tax-free status amounted to a $20 billion tax savings to Bezos's business. Baker notes, "In a state like New York, where combined state and local sales taxes average over 8.0 percent, Amazon could charge a price that was 1.0 percent below its brick and mortar competition, and still have an additional profit of 7 percent on everything it sold. That is a huge deal in an industry where profits are often just 2–3 percent of revenue." Bezos, eager to preserve this subsidy, went to work in Washington, DC, and got Republican congressman Christopher Cox and Democratic senator Ron Wyden to author the Internet Tax Freedom Act. The bill passed and was signed by President Bill Clinton on October 21, 1998. Although not barring states from imposing sales taxes on ecommerce, it does prevent any government body from imposing Internet-specific taxes. Between the passage of the ITFA and 2015, Amazon essentially managed to wipe out the local bookstore and, to some extent, the local record store from the American landscape. During those years, 2,300 independent bookstores (as well as the Borders chain) and 3,100 record stores closed their doors. The cultural critic Leon Wieseltier angrily noted, "The streets of American cities are haunted by the ghosts of bookstores and record stores, which have been destroyed by the greatest thugs in the history of the culture industry." The bitter irony is that Amazon noticed that a few independent bookstores in big cities were still thriving, so it decided to go into the bookstore business itself.

But Amazon is not alone in its avoidance of taxes. *Bloomberg Businessweek* reports, "The tactics of Google and Facebook depend on 'transfer pricing,' paper transactions among corporate subsidiaries that allow for allocating income to tax havens while attributing expenses to higher-tax countries. Such income shifting costs the U.S. government as much as $60 billion in annual revenue, according to Kimberly A. Clausing, an economics professor at Reed College in Portland, Oregon." At a time when governments around the world are putting off needed infrastructure improvements because of tax revenue shortfalls, the tax avoidance schemes of our richest technology companies are partially to blame. But of course Amazon is the leader in this regard.

According to *Publishers Weekly,* "Research conducted in March of 2014 by the Codex Group found that in the month Amazon's share of new book unit purchases was 41%, dominating 65% of all online new book units, print and digital. The company achieved that percentage by not only being the largest channel for ebooks, where it had a 67% market share in March, but also by having a commanding slice of the sale of print books online, where its share in March was estimated at 64%." Amazon's political influence has only grown as it managed to persuade the Justice Department to sue — successfully, as it turned out — Apple's tiny ebookstore for antitrust violations, thereby increasing Amazon's monopsony power. The economist Paul Krugman explains that, "in economics jargon, Amazon is not, at least so far, acting like a monopolist, a dominant seller with the power to raise prices. Instead, it is acting as a monopsonist, a dominant buyer with the power to push prices down." And this dominant power of Amazon extends around the globe. As Farhad Manjoo pointed out in the *New York Times,* "The larger Amazon gets, then, the more its rules — rather than any particular nation's — can come to be regarded as the most important regulations governing commerce."

So not only was Amazon able to confound local regulations, it was also able to use lax regulatory supervision to push the boundaries of occupational safety in its massive un-air-conditioned warehouses. Business historian Simon Head, in his book *Mindless: Why Smarter Machines Are Making Dumber Humans,* writes that Amazon is a "prime example of how in the early twenty-first century, state-of-the-art information technologies can be used to re-create the harsh, driven capitalism of the pre–New Deal era." In 2011, the *New York Times* noted a report from an Allentown, Pennsylvania, paper on conditions at the Amazon warehouse in town where temperatures inside reached 110 degrees:

> So many ambulances responded to medical assistance calls at the warehouse during a heat wave in May. . . that the retailer paid Cetronia Ambulance Corps to have paramedics and ambulances stationed outside the warehouse during several days of excess heat over the summer. About 15 people were taken to hospitals, while 20 or 30 more were treated right there, the ambulance chief told *The Call.*

For a libertarian like Bezos, the notion of installing air-conditioning in his giant warehouse goes against his notion of liberty. You don't have a right to a decent workplace; you merely have the right to pursue it, ostensibly by quitting and looking elsewhere for work. Amazon were eventually put in the position where they addressed this issue. But if you do decide to continue working in an Amazon warehouse, you will be subject to the kind of twenty-first-century monitoring that Frederick Taylor — the late nineteenth and early twentieth century management consultant and efficiency expert famous for his studies of "time and motion"— would have marveled at. All employees have a personal GPS monitor that tells them which route to walk in the warehouse to get to the next item they need to pack and exactly how many seconds they have to get there. If they do not complete the trip in the time Amazon specifies, they will get a notice and a demerit — repeated failures to meet the time targets will lead to dismissal.

These same workers know that Amazon is "looking out for them" in a different way. When they pass through the airport-style body scanners at the warehouse exit and punch the digital time clock at the end of their shifts, they are confronted by large-screen monitors that display images of several of their former coworkers in silhouette, each of whom has been fired for theft. Each image has the word *terminated* stamped across the body of the malefactor next to a rap sheet listing what he stole, when he stole it, and the value of the merchandise. Perhaps this is Amazon's libertarian version of the Puritans' public stocks. All that's missing is a scarlet *T* tattooed on their foreheads.

For all his libertarian distrust of government, Bezos has never stopped trying to use government to his advantage. Some say he bought the *Washington Post* so he could have powerful influence in the nation's capital. Certainly the purchase would not make sense on a pure return-on-investment basis. Bezos knows this because he already has taken advantage of publishers. In June of 2009 James Moroney, publisher of the *Dallas Morning News,* testified in Congress about his negotiations with Amazon over publishing the newspaper's content on the Amazon Kindle. Amazon demanded 70 percent of the subscription revenues, leaving him with 30 percent to cover the cost of creating 100 percent of the content. This, he noted, could hardly be characterized as a fair business deal.

6.

It is Peter Thiel's investment in Palantir that demonstrates a certain libertarian hypocrisy about corporate welfare. Thiel is always complaining about crony capitalism, but the initial investment in the company came from the CIA's venture capital arm, and the firm is now valued at more than $10 billion. Palantir was founded three years after the 9/11 terrorist attacks on the United States and was set up as a data-mining company that could sell its services to the CIA. Palantir's data-mining software has illuminated terror networks and figured out safe driving routes through war-torn Baghdad. It has also tracked car thieves and traced salmonella outbreaks. In one of the great ironies of history, US attorneys deployed its technology against the hedge fund SAC Capital, which was also an early investor in Palantir. Somehow, taking money from the CIA and aiding the NSA and FBI in tracking civilians does not seem contradictory to the extreme libertarian Thiel. "I felt we were drifting to a place in the U.S. we'd have a lot fewer civil liberties and no real effective protection," Thiel stated when asked why he started Palantir. The journalist Ben Tarnoff, writing in the *Guardian,* views Thiel's subsequent embrace of Donald Trump's economic plan as a rather cynical ploy to run the Palantir model at scale.

> Following this logic, what's needed is a state that bankrolls scientific research at midcentury cold war levels — without the comparatively high tax rates and social spending that accompanied it. Corporations would mine this research for profitable inventions. The public would foot the bill and ask for nothing in return.

As Thiel moves into President Trump's inner circle, we can expect the fortunes of Palantir to improve.

Born from the marriage of counterculture idealism and Defense Department funding in the 1960s, the Internet had by 2002 morphed into a vast commercial and government surveillance platform. The goal of Tim Berners-Lee and Stewart Brand — to construct a new platform for democratic communication — had been co-opted by a new cadre of libertarian *übermenschen,* a group of men who believed that they had both the brilliance and the moral fortitude to operate outside the normal strictures of law and taxes. These men believe in their "super human" qualities so deeply that they are investing millions of dollars in ventures like Thiel's Halcyon Molecular, a company that purports to "create a world free from cancer and aging." The men believe that technology will eradicate the fundamental human anxiety that is fear of death.

The futurist Jaron Lanier once described Google chief scientist Ray Kurzweil and his concept of "the singularity": the point in time when an artificially intelligent machine will be capable of autonomously building ever smarter and more powerful machines than itself. Lanier noted, "These are ideas with tremendous currency in Silicon Valley; these are guiding principles, not just amusements, for many of the most influential technologists.... All thoughts about consciousness, souls, and the like are bound up equally in faith, which suggests something remarkable: What we are seeing is a new religion, expressed through an engineering culture." Bill Gates commented, "It seems pretty egocentric while we still have malaria and TB for rich people to fund things so they can live longer."

But this is not stopping Thiel and Page in their pursuit of everlasting life. As the journalist Charlotte Lytton noted in December of 2015, "The 2045 Initiative – Dmitry Itskov's life-extension organization seeking to transfer personalities onto non-biological items and, ultimately, immortality – projected that this year could be the first in which such a system was created." I can't imagine that Thiel and Page have really thought this concept of immortality through to its logical conclusion. Obviously the treatment would be extremely expensive and only available to the very rich. For the whole of history, the less well off have at least comforted themselves with the thought that death was evenhanded – even John D. Rockefeller was not going to be able to buy off the Grim Reaper. Imagine the rage these immortality clinics would foment. The very rich, when they get to be 130 years old or more, would be so fearful of ordinary causes of death – a car accident, a plane crash, a terrorist bomb – that, having spent millions of dollars on immortality, they might be afraid to leave their mansions for fear of losing money on their investment.

I would say it takes no big leap to guess that both Peter Thiel and Larry Page truly believe that technology can deliver happiness. In a new book, *The Internet of Us: Knowing More and Understanding Less in the Age of Big Data,* Michael Patrick Lynch starts with a thought experiment: "Imagine a society where smartphones are miniaturized and hooked directly into a person's brain." Google's Larry Page is already working on this. Then Lynch takes us several generations into the future, where we have stopped learning by observation and reason and have become totally dependent on the Google Now chip in our brains. And then imagine some disaster disables the worldwide communications grid. It would be, Lynch says, as if the whole world had gone blind. Not only would we have lost the knowledge to which we once had access, we also wouldn't have the ability to learn anything new.

Bob Dylan once wrote, "To live outside the law, you must be honest." In the 1990s, the challenge set before those libertarian entrepreneurs who would go on to survive the dot-com crash of 2000 would be to live that dialectic. They created a world dominated by tech elites, with a set of rules that we are now condemned to live by. Peter Thiel declared that "competition is for losers," and the following decade would prove just who the losers would be.

CHAPTER FIVE

Digital Destruction

This renegade thing was very clear at Napster.
— Sean Parker

1.

Sean Parker, sitting in the back row of his afternoon world civilization class, was totally bored. He had stayed up most of the night before, hacking into the computers of a Fortune 500 firm that was at the heart of the American economy, and now he was having a hard time keeping his eyes open. A secretary to the principal came into the classroom and brought him a note saying that his father was waiting outside to take him to an orthodontist appointment. Sean didn't wear braces. His heart began to race. He got up with a real sense of dread and left the classroom. His father had come downstairs at five o'clock that morning to find Sean at his computer, deep into the company network, and flew into a rage. "So," Parker recalled in a 2010 *Vanity Fair* article, "he grabbed the keyboard from my hands, ripped it out of the computer, and took it upstairs. I started crying and saying, 'Dad! You don't know what you're doing! I have to log out!' But he didn't let me." Within hours the FBI traced the hack to Parker's IP address and, with assistance from his Internet service provider, raided his house. His dad, who had taught Sean computer programming when the boy was just seven years old, hauled him out of school and brought him home, where he was arrested. His father worked for the government as an oceanographer and, perhaps out of a sense of guilt over his role in Sean's hacking obsession, managed to pull some strings. Because Sean was a minor he was sentenced to two hundred hours of community service instead of jail time.

Sean performed his community service with other delinquents at a local library. In what could have been a scene from *The Breakfast Club*, he met a "punk rock princess" there to whom he lost his virginity, and he connected online through a library computer with another hacker named Shawn Fanning. Together they would invent the beginning of the end of the music business as we knew it. Their software was called Napster, and its logic was simple. Every music CD consists of a digital file that can be ripped — that is, digitally extracted — and shared on a computer network. What was needed was a way to index all those MP3 files and allow users to share them for free. Because neither Parker nor Fanning is a musician, it doesn't seem to have occurred to them the incredible damage their invention would inflict on the livings of artists for decades to come. What they knew is that offering free music was the perfect way to drive traffic to their site on the still-sleepy Internet. Parker, of course, knew they were breaking the law, and his emails admitting that users were infringing copyright were used in the lawsuit filed by the record companies that eventually shut down Napster. Before it was shut down, in July of 2001, Napster had seventy million registered users. The chart below shows the precipitous fall in music revenues from the point when Napster went online.

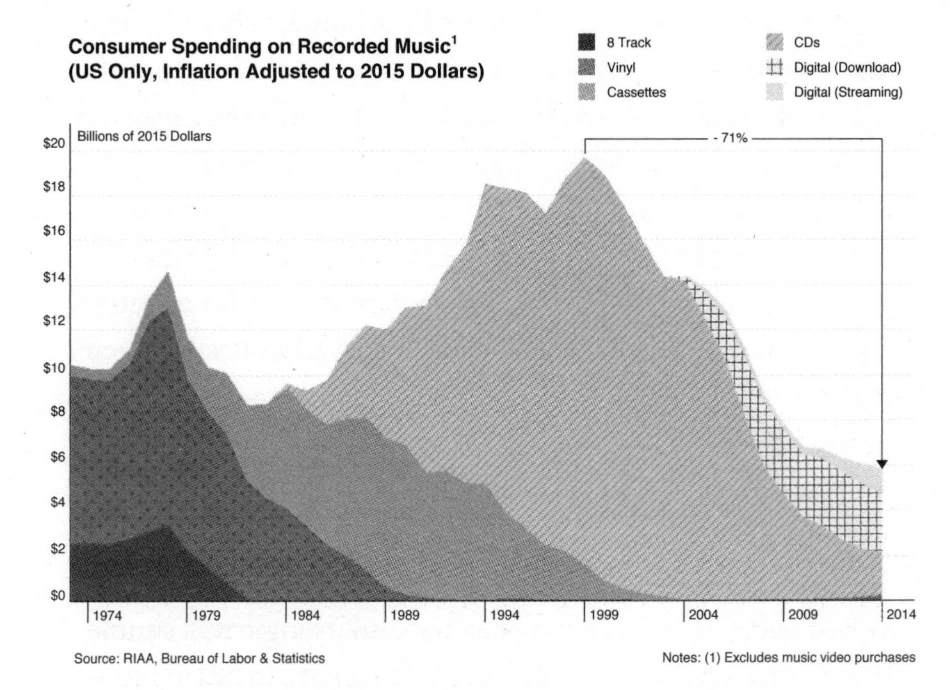

If 1999 was the high point of the music business, the onset of Napster and all the pirate sites that sprang up after it was shut down was the low point. Those sites turned the recorded music industry from a $20 billion business to a $7.5 billion business. Imagine if any other industry had been cut by two-thirds because of counterfeiting.

2.

For Sean Parker there was no downside to being a convicted corporate hacker if it also meant that he could be the inventor of the first technology to steal intellectual property. In the movie about Facebook, *The Social Network*, Parker is played by Justin Timberlake. Sean's bad-boy persona, captured perfectly by Timberlake, appealed to a young Mark Zuckerberg, who eventually appointed Sean president of Facebook. Parker served an important function, introducing Zuckerberg to Peter Thiel, who became the company's first outside investor. But Parker couldn't shake his bad-boy habits, and in 2005 he was busted on suspicion of cocaine possession on a kiteboarding trip in North Carolina during a late-night raid on a house he was renting. Once again Parker managed to avoid prosecution, but he agreed to leave Facebook. Thiel, of course, stayed loyal to his friend, hiring him at his Founders Fund venture capital firm and noting, "Sean is one of the great serial entrepreneurs of his generation, someone who is really changing the world and turning the wheel of history."

Thiel's Randian ethical framework makes it easy to cast Parker as someone who's "turning the wheels of history" despite the morally challenged projects his fame is built upon. The notion that Parker is a great disrupter cannot be disputed, but the fact that his businesses were built on copyright theft (Napster) and deep consumer surveillance (Facebook) leads us to question what exactly these attention harvesting industries create and whether they're aiding the larger culture or destroying it. Disruption of critical cultural infrastructure is only worthy if the replacement is more beneficial to the society at large than the original institution was. For instance, has the wholesale destruction of the newspaper industry (see chart below) been followed by the establishment of a more reliable source of local and global news? Or has it just resulted in noise and confusion? Thiel and Parker's Facebook currently provides 70 percent of the inbound traffic for *BuzzFeed* and the *Huffington Post*. These three sites are now the sole sources of news for many of their online users. How long before Facebook becomes the controlling force in the online journalism business?

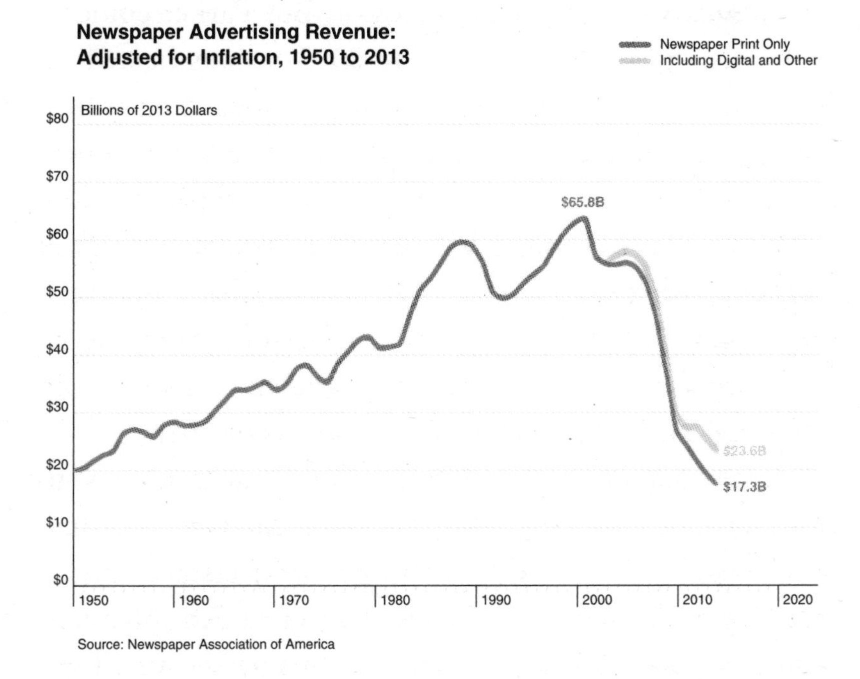

The disrupters of Silicon Valley have a very high opinion of their place in history and the role of technology in forcing change and evolving the economy. As Parker puts it, "It's technology, not business or government, that's the real driving force behind large-scale societal shifts." But as the atomic scientist Robert Oppenheimer discovered after Hiroshima, technological progress has no inherent moral arc. Parker himself explains this best in an interview with the journalist David Kirkpatrick:

> I think the best way to describe me is as an archetypal Loki character, like Joseph Campbell's Hero with a Thousand Faces. I'm like the prankster or Puck in mythology. He's not trying to cause harm, but rather to pull back the veil that masks your conventional, collectively reinforced understanding of society. This renegade thing was very clear at Napster. The point was that the emperor – the content industry – had no clothes. This all probably sounds incredibly pretentious and narcissistic.

In the same interview Parker bemoaned contemporary culture's scarcity of revolutionary thinkers like Jim Morrison or Jack Kerouac. It is a bit rich to hear the rebel venture capitalist who travels every year to the Burning Man festival in the hope of recapturing his Jim Morrison dream bemoan the tired thinking of his fellow Americans – despite the wholesale wreckage his adventures have wrought on American culture, including the very music business Morrison depended upon. For the uninitiated, Burning Man is an annual gathering that takes place in the Black Rock Desert of Nevada over Labor Day weekend. It has been described as "an experiment in community and art" influenced by ten main principles, including "radical inclusion, self-reliance, and self-expression." This libertarian Woodstock involves a lot of drug taking, naked dancing, and general madness — the perfect place for Parker, born years after the original music festival, to play out his rock-rebel fantasies.

3.

The German novelist Thomas Mann would have described Sean Parker as a bourgeois manqué, a man of passing artistic desire who is nevertheless chained to a world of suits and bureaucratic concerns. That would also describe Larry Page, founder and CEO of Google, a young man who played the saxophone in high school and studied music composition. Page tried to invent a music synthesizer while at the University of Michigan, but was unable to create the real-time software that the device needed. Like Parker (and perhaps his fellow music lover Steve Jobs), Page didn't have the talent to make it as a professional musician, so he applied for the master's program in computer science at Stanford University and enrolled in the PhD program there in 1995.

The most important element Page seems to have derived from his music education is the importance of speed. "In some sense I feel like music training led to the high-speed legacy of Google for me," Page said during an interview with *Fortune*. "In music you're very cognizant of time. Time is like the primary thing." Larry Page probably never listened to early Louis Armstrong records, in which the notion of time is reinvented. Armstrong threw away the classical meanings of time, which he had learned in the marching bands of New Orleans, and replaced it with a slippery push-pull syncopation that would come to define the idea of swing. Perhaps we can get a small insight from this into the culture clash between the nerds and the artists. The time clock in a computer is unforgiving, but a great musician often plays behind the beat. Once computers became ubiquitous in music, then the computer clock (e.g., disco's 120 beats per minute) brought Page's idea of time into popular music.

Like Sean Parker, Page liked Burning Man, saying, "[It's] an environment where people can try new things. I think as technologists we should have some safe places where we can try out new things and figure out the effect on society. What's the effect on people, without having to deploy it to the whole world?" In fact, Page and his partner Sergey Brin were so entranced with Burning Man that they cited Eric Schmidt's attendance there as one of the major reasons they agreed to hire him as CEO after the Google board decided that Page and Brin needed "adult supervision." I suppose Parker and Page both embrace Burning Man because for a long weekend it approximates the "free cities" model they embrace – in which polities are privately owned and unregulated by states – an ideal way for capitalists to avoid the "mob mentality" of democracy.

This need for control also played itself out when Google went public. Brin and Page set up a two-class stock structure (mimicking monopoly cable firms such as Comcast) in which their own shares had ten times the voting power of the shares offered to the public. As Page explained in his first letter to shareholders, "New investors will fully share in Google's long term economic future but will have little ability to influence its strategic decisions through their voting rights." Page and Brin set out to deliberately "disrupt" the classic idea of a public corporation. The very idea of shareholder democracy was anathema to them. They wanted all the advantages of the public's investment capital but none of the disadvantages of having to answer to shareholders.

4.

The founding of Google is one of the key moments in the history of the Internet, and so it is important to take some time to understand its philosophical principles. As Ken Auletta pointed out in his book *Googled: The End of the World as We Know It,* Page and Brin never asked for permission to copy the entire World Wide Web onto their servers and then index it. Ayn Rand's famous quotation "Who will stop me?" seems to be the founding principle of Google. Page's constant assurances in his initial shareholder letter that everyone should trust his and Brin's good intentions are critical to this Randian mind-set: "Don't be evil. We believe strongly that in the long term, we will be better served – as shareholders and in all other ways – by a company that does good things for the world even if we forgo some short term gains. This is an important aspect of our culture and is broadly shared within the company." This naïveté, this barely disguised will to power, this dialectic – Google will do whatever it wants without asking permission, and the results will be so awesome that no one will complain — stands at the heart of the company's success. Gmail and Google Street View are two examples. The trade-off Google offered customers — allow Google to scan all your email (so they can place customized advertising on it) in return for a gigabyte of free storage — is not something users would have accepted if the company had asked permission up front. The offer was free email with unlimited storage. The fine print said, in effect, "You give us permission to read your mail in order to sell you stuff." In addition, customers were not asked if the Google Street View camera could take a picture of their front yards and match it to their addresses.

But Page seems unconscious of the irony of the dialectic. He has stated, "For me, privacy and security are a really important thing. We think about it in terms of both things, and I think you can't have privacy without security." But no one has done more to eliminate the notion of privacy than Larry Page. For the first two years of Gmail, Page refused to put a Delete button on the service, because Google's ability to profile you by preserving your correspondence was more important than your ability to eliminate embarrassing parts of your past. The data smog you leave behind was also so crucial to "personalized search" that Google chairman Eric Schmidt told *The Atlantic*, "We don't need you to type at all. We know where you are. We know where you've been. We can more or less know what you're thinking about." In Europe, Google continues to challenge "the right to be forgotten" — customers' ability to eliminate false articles written about them from Google's search engine.

Google took a similar "don't ask permission" tack when Brin decided to digitize all the world's books. As he told Auletta, if they had asked authors and publishers, Google "might not have done the project." This same dynamic plays itself out on Google's YouTube platform, where the company has somehow managed to make it the responsibility of the content owner to police the site for copyright infringement. YouTube watch time is growing 60 percent year over year, and revenues could reach $12 billion in 2017. Because of the "don't ask permission" policy, every single tune in the world is available on YouTube as a simple audio file (most of them posted by users). So as the following chart shows, YouTube is the largest streaming music site in the world, with a 52 percent market share, even though it pays only 13 percent of the streaming music revenues that the music business does.

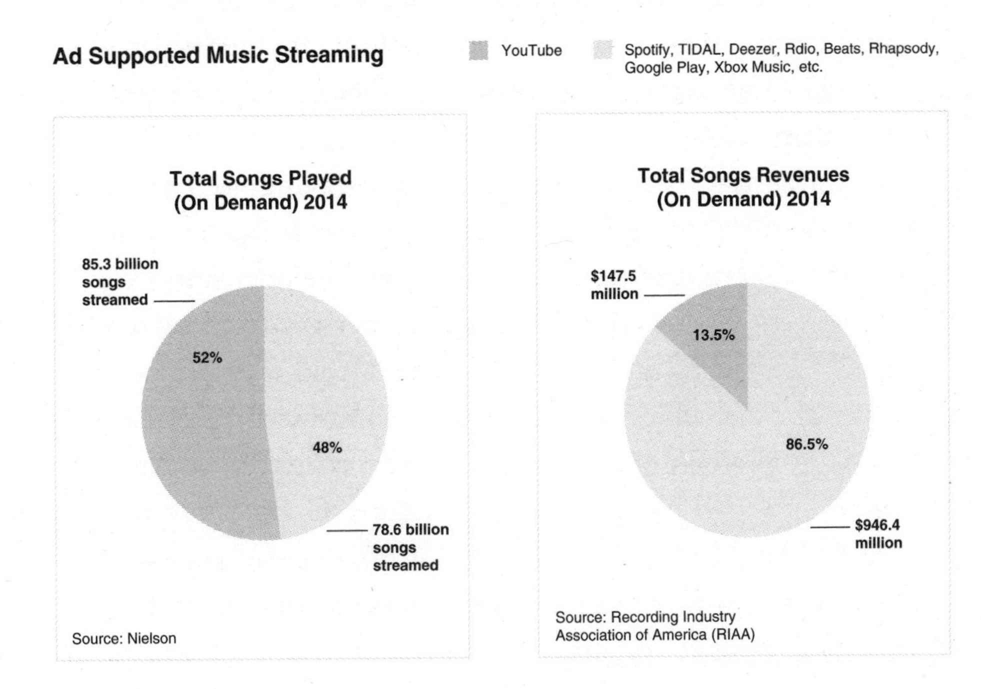

The cofounder of YouTube, Chad Hurley, was a PayPal alumnus, schooled in Peter Thiel's philosophy. He built his company on the same "don't ask permission" ethic that Larry Page had embraced. Emails released during one of the early copyright infringement lawsuits against the company make it clear that Hurley knew he was building his business on illegal conduct. In a June 15, 2005, email to his cofounders, Hurley writes, "So, a way to avoid the copyright bastards might be to remove the 'No copyrighted or obscene material' line and let the users moderate the videos themselves. Legally, this will probably be better for us, as we'll make the case we can review all videos and tell them if they're concerned they have the tools to do it themselves."

Here Hurley shows an early appreciation of the "safe harbor" provisions of the Digital Millennium Copyright Act (DMCA), which Bill Clinton had signed into law just weeks after Google went live on the Web. The statute protected online service providers (OSPs) such as Google and YouTube from copyright infringement prosecution provided that the OSP not have the requisite level of knowledge of the infringing activity…not receive a financial benefit directly attributable to the infringing activity [and] upon receiving proper notification of claimed infringement…expeditiously take down or block access to the material. Since Hurley's 2005 email, this has been YouTube's strategy: pretend not to know there is infringing material being uploaded by users and take down the content when notified by the copyright owner. But this of course neglects one crucial provision of the DMCA – does YouTube receive financial benefit directly attributable to the presence of infringing content on the site? The answer, of course, is yes: in fact you could argue that YouTube achieved success in a crowded field precisely because of its laxity toward pirated content. Competitors such as Yahoo and RealNetworks were thinking they were in a boxing match, where there were rules, while YouTube was performing in a professional wrestling match, where there are no rules.

In another email exchange from 2005 made public in the lawsuit, when full-length movies were being posted on YouTube, Steve Chen, a cofounder of the company, wrote to his colleagues Hurley and Jawed Karim, "Steal it!," and Chad Hurley responded: "Hmm, steal the movies?" Steve Chen replied: "We have to keep in mind that we need to attract traffic. How much traffic will we get from personal videos? Remember, the only reason why our traffic surged was due to a video of this type.... viral videos will tend to be THOSE type of videos."

A year after all these emails were written, Google acquired YouTube for $1.65 billion in Google stock. But the predatory conduct did not stop. In the *Viacom International, Inc. v. YouTube, Inc.* lawsuit discovery, a June 8, 2006, presentation outlining YouTube's content strategy was uncovered. It was sent by Google senior vice president of product management Jonathan Rosenberg to Google CEO Eric Schmidt and Google cofounders Larry Page and Sergey Brin, and it said, in part, that Google must "pressure premium content providers to change their model towards free[;] Adopt 'or else' stance re prosecution of copyright infringement elsewhere[;] Set up 'play first, deal later' around 'hot content.'" It also said that Google "may be able to coax or force access to viral premium content," "threaten a change in copyright policy," and "use threat to get deal sign-up."

Of course Google wasn't quite as cavalier with its own intellectual property. As their IPO prospectus had warned, "Our patents, trademarks, trade secrets, copyrights and all of our other intellectual property rights are important assets for us.... Any significant impairment to our intellectual property rights could harm our business or our ability to compete."

5.

What we have been witnessing since 2005 is a massive reallocation of revenue from creators of content to owners of platforms. You have seen the dramatically dropping revenues

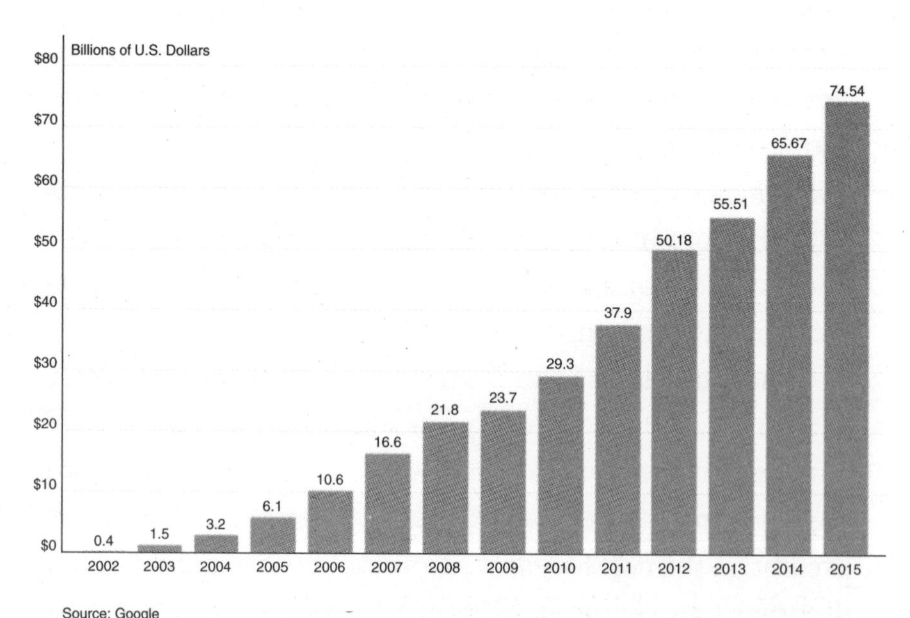

in the music and newspaper businesses. Contrast that with this chart of Google revenue, an upward trend mirrored in the revenue growth of Amazon, Apple, and Facebook.

More people than ever are listening to music, reading books, and watching movies, but the revenue flowing to the creators of that content is decreasing while the revenue flowing to the big four platforms is increasing. Each of these platforms presents a different challenge for creators. Google and YouTube are ad-supported "free riders" driven by a permissionless philosophy. Facebook, with its libertarian financier's roots, takes much of the same stance toward content and advertising, but there are signs that its CEO has real ethical questions about where the company is going. Amazon, whose founder, Jeff Bezos, embraces the libertarian creed but has not taken the "don't ask permission" route, has instead opened a new front: a relentless push to lower prices and commoditize content (especially books), which presents a different danger. And then there is Apple, the dissenter from the libertarian creed. Both Steve Jobs and Tim Cook have been real allies to the content community, and their stance against the surveillance-marketing model that is at the core of Google's and Facebook's businesses – i.e., their support of ad blockers – puts them in direct opposition to the dominant search and social platforms.

The tech elite's jealous guarding of its own monopoly platforms is built upon blatant disregard for the artist's intellectual property, a fact that came home to me in April of 2012 when I debated Alexis Ohanian, the founder of Reddit, at the *Fast Company* Innovation Uncensored event. Reddit is an online bulletin board where users can post links on almost any topic (called subreddits). Various subreddits have been devoted to pornography, neo-Nazi propaganda, white power, Gamergate – all in the name of free speech. Since Reddit's purchase by Condé Nast in 2006, the site has tried to rein in the most outrageous subreddits. But the task of reining in the ultralibertarian Reddit community became too much, and in the fall of 2011 Condé Nast sold off a large share of Reddit to a group led by Sam Altman, Peter Thiel, and Marc Andreessen. At the debate, Ohanian proudly mentioned his personal consumption of "free music and movies" available on the Internet, going so far as to say that musicians such as The Band need to earn their money from touring. From his point of view, Levon Helm had no right to make money from old recordings. Feeling that he may not have made a good argument at the conference, Ohanian wrote an open letter to me the next day, which *Fast Company* published.

> Thanks for debating me this evening at Fast Company's Innovation Uncensored conference here in NY! Like I said on stage, I wanted to offer a solution to help make right what the music industry did to members of The Band...I'm hopeful that innovations like the ones I discussed tonight and the others that are being worked on by entrepreneurs right now will continue to do right by artists and cut out those who'd mistreat them...Like I said on stage, it would be an honor to gather members of The Band together to produce one more album with unreleased content or something to honor Levon Helm – really any

kind of creative project they'd like to produce – (this time funded on kickstarter) and we'll gladly launch it on the IAMA section of reddit.

I replied in an open letter to him.

Dear Alexis,

Last week at our debate, I talked about the essential unfairness that my friend and colleague Levon Helm had to continue to tour at the age of 70 with throat cancer in order to pay his medical bills. On Thursday, Levon died and I am filled with unbelievable sadness. I am sad not just for Levon's wife and daughter, but sad that you could be so condescending to offer "to make right what the music industry did to the members of The Band." It wasn't the music industry that created Levon's plight; it was people like you celebrating Pirate Bay and Kim Dotcom – bloodsuckers who made millions off the hard work of musicians and filmmakers.

You were so proud during the debate to raise your hand as one of those who had downloaded "free music and free movies." But it's just your selfish decision that those tunes were free. It wasn't Levon's decision. In fact, for many years after The Band stopped recording, Levon made a good living off of the record royalties of The Band's catalog. But no more.

So what is your solution – charity. You want to give every great artist a virtual begging bowl with Kickstarter.

> But Levon never wanted the charity of the Reddit community or the Kickstarter community. He just wanted to earn an honest living off the great work of a lifetime.
>
> You are so clueless as to offer to get The Band back together for a charity concert, unaware that three of the five members are dead. Take your charity and shove it. Just let us get paid for our work and stop deciding that you can unilaterally make it free.

Ohanian did not respond.

The Internet was supposed to be a boon for artists. It was supposed to eliminate the "gatekeepers" — the big studios and record companies that decide which movies and music get widespread distribution. But a single statistic from the music industry puts a lie to this notion. When I started in the entertainment business, we talked about the Pareto curve, otherwise known as the 80-20 rule — the idea is that a movie or record company derives 80 percent of its revenue from 20 percent of its products. In other words, one in five movies becomes a big hit. In 2015, in the music business, 80 percent of the revenue came from 1 percent of the products. So Jay Z, Taylor Swift, and a few others got really rich, and most musicians made little or no money from their recordings. It could be that the very nature of search engines, which push the most popular items to the top of the search results, is reinforcing this winner-takes-all economy. Is a tiny percentage of constantly circulating material the rich diversity the Internet has brought us?

There is a deeper problem as music moves to an all-streaming format. Because services such as Spotify and

YouTube spew out terabytes of data that record companies can analyze, the whole business is becoming ever more data-driven. Data is fabulous for showing people what's already popular, but it's terrible for pointing the way toward art – great breakthroughs arising from something that has never been done before. Reliance on data leaves both Hollywood and the music business stuck in a remake-and-sequel culture – artifacts that data agrees will be successful. As the writer Kurt Andersen pointed out in *Vanity Fair,* "Even as technological and scientific leaps have continued to revolutionize life, popular style has been stuck on repeat, consuming the past instead of creating the new." The death of David Bowie leads me to wonder if an artist like he could ever have a career in today's music business. He didn't make hits; he made complex works that took a while to understand. And even Bowie knew he was lucky to have started his career in a predigital age, before music had become a commodity, telling a reporter in 2002, "Music itself is going to become like running water or electricity. So it's like, just take advantage of these last few years because none of this is ever going to happen again." When you think that Bob Dylan's first album sold four thousand copies in two years, you realize that today his contract would never have been renewed.

The techno-utopians like Alexis Ohanian told us that the Internet would "kill all the gatekeepers." But what's really happened is that a new set of gatekeepers – Google and Facebook – has replaced the old. Google's market capitalization is $532 billion. Time Warner's is $61 billion. The balance of power in the world of entertainment has shifted to monopoly platforms. To understand how that happened, we need to look at the nature of monopoly capitalism and how an old and largely discredited form of robber-baron capitalism took on a new form in the digital age.

CHAPTER SIX

Monopoly in the Digital Age

Competition is for losers.
— Peter Thiel

1.

Robert Bork did more than any individual in the twentieth century to embed the libertarian free-market principles of Ayn Rand and Milton Friedman into the heart of the American economic and judicial system. Professor Bork's class on antitrust law at Yale Law School was usually packed. In 1971, he had taught Bill Clinton and his soon-to-be wife, Hillary, as well as Robert Reich, Clarence Thomas, and Richard Blumenthal. Bork was tall, with very curly brown hair, big tortoiseshell glasses, and a beard that made him look like a cross between an Amish farmer and a rabbi. He was partial to dark suits and Brooks

Brothers button-down shirts, and his one concession to fashion was his tendency to favor paisley ties.

The theory of antitrust Bork taught (wags called it a course in pro-trust law) came out of his original thinking on the nature of business regulation. Reich recalls Bork's class as being quite contentious: "I didn't know how much his ideas would influence antitrust policy in future years. At the time they seemed absurd to me because they looked at only one dimension of public policy – consumer welfare – and disregarded all of the dynamic effects of concentration, such as political power or predation." In March of 1973, Bork would put his classroom concepts to work as solicitor general of the United States during the Richard Nixon and, later, the Gerald Ford administrations. One measure of how little Bork thought of the Sherman Act (our main antitrust statute) is that its filings dropped substantially during Bork's term. When Bork left the Justice Department he published *The Antitrust Paradox: A Policy at War with Itself,* which has shaped antitrust law ever since, mostly by focusing the discipline on efficiency and articulating its goal as "consumer welfare."

In other words, Bork argued that the sole matter that should concern regulators was whether prices to consumers were falling. From Bork's point of view, if Walmart ended up as the only general retailer in the country, as long as prices continued to fall, this would benefit consumer welfare. In 1984, while Bork was a judge in the United States Court of Appeals for the District of Columbia Circuit, I was a vice president for mergers and acquisitions at Merrill Lynch's investment banking division. We helped

Gulf Oil merge into Standard Oil of California, thus creating Chevron and cutting the number of major oil companies from seven to six. During the whole process, the question of antitrust problems never came up.

Bork centered his interpretation of antitrust laws in the Chicago school libertarian economic theories of Milton Friedman. Bork came to the University of Chicago in the late 1940s as a true believer in Roosevelt's New Deal politics and started dating a similarly liberal undergraduate, Claire Davidson. But as he began to get attracted to the intersection of economics and law, the conservative professors who dominated the Chicago faculty discouraged any belief that government could or should play a role in regulating business. The free market needed to function, unencumbered by government regulation. Whether they were motivated by a desire to advance in academia or a true change in belief, both Bork and Davidson (whom he married in 1952) came under the sway of the Chicago school and subscribed to its theories. From that point on he fought government regulation and continued to fight it for the rest of his life. As the *New York Times* noted in his obituary, this stance cost him a seat on the Supreme Court: "He also wrote a fateful article for The New Republic in 1963 – one that played a key role in his 1987 confirmation defeat – condemning the public accommodation sections of the proposed 1964 Civil Rights Act aimed at integrating restaurants, hotels and other businesses. Mr. Bork said he had no objection to racial integration but feared that government coercion of private behavior threatened freedom."

Google, Amazon, and Facebook are all monopolies

that could be prosecuted under antitrust statutes if it hadn't been for Robert Bork. From the Ford administration all the way through to the Obama White House, Bork's principles as expressed in *The Antitrust Paradox*, encouraging mergers and calling for less regulation, have ruled the antitrust division of the Justice Department. Just a few years after Bork published his book, Reagan's soon to be appointed head of the antitrust division, William Baxter, told the *New York Times* that he would "pursue an antitrust policy based on efficiency considerations." As Barry Lynn points out in his book *Cornered: The New Monopoly Capitalism and the Economics of Destruction*, opposition to Baxter's policy came from senators of both parties who argued that the Sherman Act, in which the word *consumer* never appears, was meant "not to lower prices but to protect the independent entrepreneur and to prevent a few among us from using our political and economic institutions to concentrate power in their own hands."

2.

Why do antitrust laws matter, besides occasionally helping the little guy compete against corporate giants? To answer that question, we need to know that the fear of monopoly goes back to the founders of the American republic and the epic battle between Alexander Hamilton and Thomas Jefferson. Hamilton, of course, represented the commercial power of New York and the financiers who held the bonds that financed the revolution. Jefferson, the man of the people, had been sent to France in 1784 and became our minister to the country that had also helped finance the revolution. While he was there, the meetings in Philadelphia to draw up the Constitution were going on, and James Madison had to represent Jefferson's point of view in the formation of the document. Jefferson's major objection was that it included no bill of rights. He wrote Madison from Paris, having read the first draft.

> I will now add what I do not like. First, the omission of a bill of rights, providing clearly, and without the aid of sophism, for freedom of religion, freedom of the press, protection against standing armies, **restriction against monopolies** [emphasis added], the eternal and unremitting force of the habeas corpus laws, and trials by jury in all matters of fact triable by the laws of the land, and not by the laws of Nations.

Eventually, with Madison's help, he got the assembled delegates to write the Bill of Rights, but the one clause Hamilton's Federalists fought fiercely was "restriction of monopolies." From his perch in Europe, Jefferson had first-hand knowledge of the corrupting power employed by monopolies such as the British East India Company. The company, which had an absolute monopoly on English trade in India and China, became so rich and powerful that in 1708 it loaned the nearly bankrupt British treasury three million pounds in return for extending its monopoly for a much longer period. Jefferson also observed that the monopoly rents (the extraction of extra profit from a monopoly) the officers of the company enjoyed allowed them to come home to England and establish sprawling estates and businesses and to obtain political power. The company developed a lobby in Parliament that was so powerful that they could write legislation pledging British military might to protect their private trade routes. Jefferson had also read of the Great Bengal Famine of 1770, which resulted in the deaths of ten million people. The British East India Company forced Bengal farmers to grow opium – which the company intended to export to China – instead of food crops, resulting in a shortage of grain for the local inhabitants. Jefferson saw the havoc that unrestrained monopoly could deliver.

Hamilton wanted to have it both ways, believing that capital should be free to influence politics but that politics should not be allowed to influence capital. He had already founded the Bank of New York in 1784 and was determined to found a national bank, 20 percent of whose shares would be owned by the US government. The remaining 80 percent would be owned by his friends. It would be a monopoly. In February of 1791, over the objections of Jefferson, Madison, and their allies, Hamilton got the bill passed through Congress. Jefferson, who was by then secretary of state, appealed to George Washington to veto the bill, but the president gave in to Hamilton, his secretary of the treasury. Thus the stage was set for American business to be ruled by giant corporations.

By the end of the nineteenth century, men like John D.

Rockefeller and J. P. Morgan ruled combinations of companies called trusts. Faced with the real threat of monopoly, Congress passed the Sherman Act in 1890, which specified fines and imprisonment for anyone "who shall monopolize, or attempt to monopolize, or combine or conspire with any other person or persons, to monopolize any part of the trade or commerce among the several States, or with foreign nations." President Theodore Roosevelt used this law in his first term to break up Rockefeller's Standard Oil trust and Morgan's Northern Securities trust. Here are four of Roosevelt's clear statements on the dangers of monopoly:

> The great corporations which we have grown to speak of rather loosely as trusts are the creatures of the State, and the State not only has the right to control them, but it is duty bound to control them wherever the need of such control is shown.
>
> The way out lies, not in attempting to prevent such combinations, but in completely controlling them in the interest of the public welfare.
>
> Corporate expenditures for political purposes... have supplied one of the principal sources of corruption in our political affairs.
>
> The absence of effective State, and, especially, national, restraint upon unfair money-getting has tended to create a small class of enormously wealthy and economically powerful men, whose chief object is to hold and increase their power. The prime need is to change the conditions which enable these men to accumulate power which it is not for the general welfare that they should hold or exercise.

3.

Even though it was Ronald Reagan who created the break with historical antitrust policy, the blame cannot be placed on Republican administrations alone. As Barry Lynn points out, Bill Clinton's "attitudes towards monopolization were even more favorable than those of Reagan or George H. W. Bush." And even though Clinton and Gore ran in the 1992 presidential campaign as opponents of media monopolization, Lynn says, "their decision to allow the consolidation of U.S. media companies that had begun under Reagan to continue...cut the number of big firms from more than fifty to six." And of course since Clinton, the libertarian theories of Robert Bork have continued to rule in the George W. Bush and Obama administrations and were reflected in a key 2004 opinion by Supreme Court justice Antonin Scalia: "The mere possession of monopoly power, and the concomitant charging of monopoly prices, is not only not unlawful; it is an important element of the free-market system." But Scalia is not alone. A study conducted by Lee Epstein at the University of Southern California, William M. Landes of the University of Chicago Law School, and Judge Richard A. Posner of the federal appeals court in Chicago found that five of the ten most business-friendly justices since 1946 sat on the Supreme Court in the last term that Scalia was on the court.

But how could it be that such a profound change in policy has escaped the notice of most Americans? Perhaps if you own a bar and only have one beer supplier (as the merger of the two beer giants A-B InBev and SABMiller has been approved), then maybe the consequences of monopoly will reach into your life, but for most of us it goes unnoticed, even though its effects on our well-being are profound. As economist Yelena Larkin and her colleagues point out, increasing consolidation has reduced the number of US publicly traded firms by more than 50 percent (in half of US industries): "This decline in the number of firms has been so dramatic that the number of firms these days is lower than it was in the early 1970's, when the real gross domestic product in the U.S. was one third of what it is today."

Two senior Obama economic advisers, Peter Orszag and Jason Furman, published a paper entitled "A Firm-Level Perspective on the Role of Rents in the Rise in Inequality," which makes the argument that the rise in "super-normal returns on capital" at firms with limited competition is leading to a rise in economic inequality. They describe these firms as "rent seeking."

> Economic rents are the return to a factor of production in excess of what would be needed to keep it in the market.... For example, capital can extract rents by engaging in anti-competitive behavior to earn revenues well in excess of opportunity cost.... Moreover, labor market structure can lead to some elements of monopsony in certain industries, slanting the division of this labor contract rent towards the firm.

The classic example of rent seeking we all deal with is our cable service, which costs more than it should and is less responsive than it should be because its local monopoly allows the company to make a better "deal" for itself. It is probably fair to ask whether Google, Facebook, and Amazon are rent-seeking firms — i.e., whether they monopolize the resources they use. Does the fact that Amazon can deny a publisher access to its vast customer base allow it to extract rents from publishers in excess of what it would be able to extract if there were multiple large sellers in the ebook market? Yes. Can Facebook and Google extract monopoly rents in excess of normal market prices from advertisers in return for the targeted access to their billions of users? Former assistant attorney general Thomas Barnett answered that question in his September 2011 testimony before the Senate antitrust subcommittee: "First of all, remember they [Google] are an advertising company. They made $30 billion last year [$60 billion in 2015] in advertising. And given that they're dominant in advertising, a good portion of that is already monopoly rents."

Peter Orszag agrees, telling a Sydney, Australia, audience that Facebook and Google are "monopolies that are using our personal information without paying us and extracting a monopoly rent by selling ads based on that personal information." So for Orszag and Furman, inequality has been rising between firms, not within them. Yes, CEOs still make more than workers, but Google, Amazon, and Facebook workers, with their extraordinary stock options, do far better than their peers in other industries. Tech giants such as Google and Facebook may have more to do with economic inequality than we realize. "It may be that we have dramatically mischaracterized a lot of what's driving increasing inequality in the US," Orszag told the Sydney conference.

4.

The tightening monopolization of US industry is rendering America an oligarchy, with profound ramifications for our political system. After examining differences in public opinion across income groups on a wide variety of issues, the political scientists Martin Gilens of Princeton University and Benjamin Page of Northwestern University found that the preferences of a small number of corporations and the very rich had a huge impact on policy decisions while the views of middle-income and poor Americans had almost none. Gilens and Page write, "Our analyses suggest that majorities of the American public actually have little influence over the policies our government adopts." The sociologist C. Wright Mills envisioned this in the mid-1950s in his classic book *The Power Elite*. He wrote, "The long-time tendency of business and government to become more intricately and deeply involved with each other has... reached a new point of explicitness. The two cannot now be seen clearly as two distinct worlds." Mills could hardly have imagined the kind of power business exerts in the post–Citizens United world of billion-dollar political campaigns.

But it is in the world of digital media that the monopoly trend is most pronounced. As the *New Yorker*'s technology writer Om Malik put it, "Most competition in Silicon

Valley now heads toward there being one monopolistic winner." Part of this is the network effect that Peter Thiel felt was so crucial, but part of it is also the unique nature of the Internet's architecture. As Barry Lynn points out, "The entirely new fact is that the monopolist in the digital world enjoys a power that the monopolist in the physical world does not. This is the ability [not only] to isolate producers one from another and discriminate among them, but also to isolate consumers from one another and discriminate among them." In antitrust law, an HHI score – according to the Herfindahl-Hirschman Index, a commonly accepted measure of market concentration – is calculated by squaring the market share of each firm competing in a given market and then adding the resulting numbers. The antitrust agencies generally consider markets in which the HHI is between 1,500 and 2,500 to be moderately concentrated; markets in which the HHI is in excess of 2,500 are highly concentrated. The HHI in the Internet search market is 7,402. Off the charts.

Moreover, Amazon is nearly as dominant in its market as Google is in the search-engine market. The *New York Times* reported on Amazon's willingness to discriminate among publishers who were not willing to knuckle under to their terms:

> Among Amazon's tactics against Hachette, some of which it has been employing for months, are charging more for its books and suggesting that readers might enjoy instead a book from another author. If customers for some reason persist and buy a Hachette book anyway, Amazon is saying it

> will take weeks to deliver it. The scorched-earth tactics arose out of failed contract negotiations. Amazon was seeking better terms, Hachette was balking, so Amazon began cutting it off. Writers from Malcolm Gladwell to J. D. Salinger are affected, although some Hachette authors were unscathed.

Amazon has a near-monopoly position in the distribution of ebooks. The supreme irony is that government regulators are so clueless about the effects of monopoly that they brought an antitrust case against Apple in 2012 when Amazon had about 60 percent of the ebook market and Apple was a relatively minor player. But beyond books, Amazon captures fifty-one cents of every dollar Americans spend in online commerce.

It wasn't supposed to be this way. The Web's supposed low barriers to entry should have allowed a very competitive landscape, but it never happened. In search we have a monopolist in Google. In smartphone operating systems we have a duopoly in Apple and Google. And we soon might have a duopoly in home broadband service between Comcast and Time Warner (now called Spectrum). And certainly AT&T and Verizon constitute a duopoly in mobile phone service. It turns out the Internet is very good at creating winner-takes-all scenarios.

The growth of monopoly creates a system that does not function the way classical economists believe market economies should. As the digital economy becomes a large component of our GDP and companies like Google, Apple, Amazon, Comcast, Verizon, and AT&T dominate the Fortune 100, a reexamination of some of the deregulatory nostrums of the Reagan era is in order. As Peter Thiel explains in *Zero to One,* the profit margins of true monopolies are extraordinary: "Google brought in $50 billion in 2012 (versus $160 billion for the whole airline industry), but it kept 21% of those revenues as profits – more than 100 times the airline industry profit margin that year. Google makes so much money it is now worth three times more than every U.S. airline combined." The problem is that the enormous productivity of these companies, coupled with their oligopolistic pricing, generates a huge and growing surplus of cash that goes beyond the capacity of the economy to absorb through the normal channels of consumption and investment. This is why Apple has $150 billion in cash on its balance sheet and Google has $75 billion. These enterprises cannot find sufficient opportunities to reinvest their cash because there is already overcapacity in many areas and because they are so productive that they are not creating new jobs and finding new consumers who might buy their products. As former treasury secretary Lawrence Summers has put it, "Lack of demand creates lack of supply." Instead of making investments that could create new jobs, firms are now using their cash to buy back stock, which only increases economic inequality.

5.

The Harvard Business School guru Clayton Christensen (*The Innovator's Dilemma: When New Technologies Cause Great Firms to Fail*) argues, "Financial markets – and companies themselves — use assessment metrics that make innovations that eliminate jobs more attractive than those that create jobs." Whereas the return on "efficiency innovations" is relatively quick, the more important "market-creating innovations," which create entirely new industries that produce jobs, have a long time in which to return the investment. Even Silicon Valley heroes such as Elon Musk and his Tesla car are merely producing what Christensen calls "performance-improving innovations [that] replace old products with new and better models. They generally create few jobs because they're substitutive: When customers buy the new product, they usually don't buy the old product."

While economists of such different political affiliations as Paul Krugman, Larry Summers, and Tyler Cowen all have written extensively about the cause of the joblessness and "secular stagnation" in the US economy that has endured since 2000, they never examine the role that monopoly capitalism might play in this crisis. If the rise of monopoly can be seen as a cause of economic stagnation, why has it endured? Because, as Peter Thiel points out in his book, "whereas a competitive firm must sell at a market price, a monopoly owns its market, so it can set its own prices. Since it has no competition, it produces at the quantity and price combination that maximizes its profits." The Bork rule, which looks only at consumer prices, allows these digital monopolies to prosper, and it is only under such a rule that a company like Google, with an 85 percent market share in its core business, would not be subject to prosecution. It is only under the Bork rule that Amazon, with

70 percent of the ebook business, would escape judgment. And of course the Bork rule makes no provision for monopsonies like Amazon. The effect of Amazon's monopsony in the book business is to constantly force authors and publishers to work for less money. In Bork's view, as long as the customer gets lower prices, society should not care that writers cannot make a living, that independent bookstores go out of business, and that publishers die.

It may well be that Alexander Hamilton's vision of an American government controlled by the financial elites has been the norm – with the exception of the two Roosevelt presidencies. FDR grasped how much of an outlier he was when he wrote to Edward M. House, President Wilson's close adviser, "The real truth of the matter is, as you and I know, that a financial element in the larger centers has owned the government ever since the days of Andrew Jackson." But it would take Google to really perfect that model.

CHAPTER SEVEN

Google and the Risk of Regulatory Capture

Monopolists lie to protect themselves.
— Peter Thiel

1.

How have monopolies escaped regulation? Like its two peers, Facebook and Amazon, Google has used the tools of political lobbying and public relations to cement its unique market power. Regulation is at the mercy of politics, and as long as the free-market gospel that regulation inhibits growth is championed in Washington, monopolies will be free to proliferate. As the largest monopoly in the world, Google maintains the stance that governments are living in a twentieth-century world in which Google represents twenty-first-century business practices. The implied threat is that "regulators don't get it," even if those regulators are kept in check by Chicago school economic and legal doctrine. To maintain their power, Google, Facebook, and

Amazon not only have to be conscious of Justice Department strategy, they also have to keep a close watch on the Federal Trade Commission, which regulates advertising and ecommerce, as well as the Federal Communications Commission, which manages the rules of the Internet.

Google is the largest company in America (ranked by market capitalization). It controls five of the top six billion-user universal Web platforms — search, video, mobile, maps, and browser — and it leads in thirteen of the top fourteen commercial functions of the Internet, according to Scott Cleland at the consulting firm Precursor, LLC. As Peter Thiel points out, companies like Google "lie to protect themselves. They know that bragging about their great monopoly invites being audited, scrutinized and attacked. Since they very much want their monopoly profits to continue unmolested, they tend to do whatever they can to conceal their monopoly — usually by exaggerating the power of

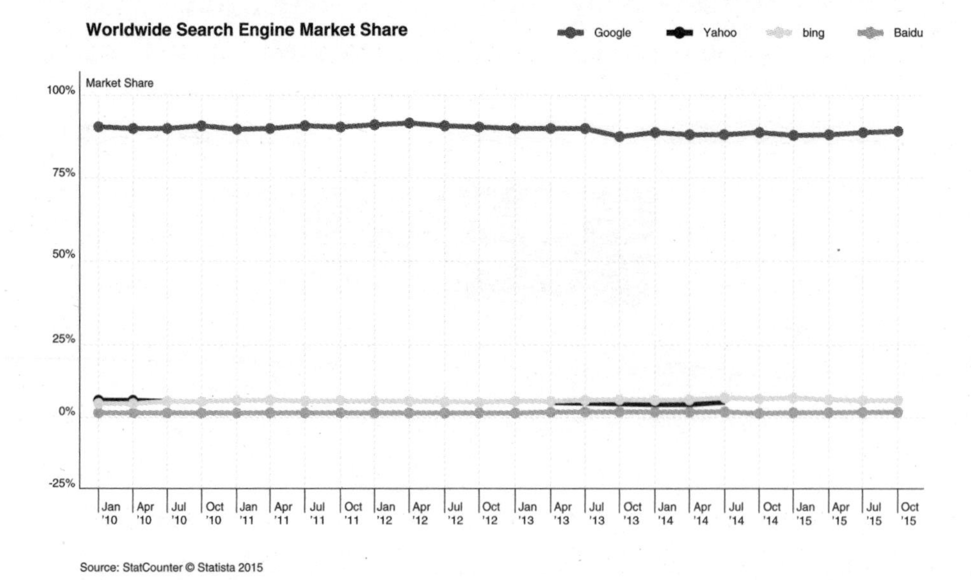

their (nonexistent) competition." This chart provides a look at the company's (nonexistent) competition.

Google's winner-takes-all success is not just a factor of its technological superiority. Google spends a lot of money to make sure its political influence in Washington is felt in both the executive and legislative branches of government. The company spends more than $15 million per year on direct lobbying: on a par with the defense contractor Boeing. But Google has a platform to reach the public that is far more powerful than anything Boeing could deploy. On January 17, 2012, the film and music industries backed the Stop Online Piracy Act (SOPA): a proposed bill that aimed to crack down on copyright infringement by restricting access to sites that host or facilitate the trading of pirated content. The bill specifically targeted search engines such as Google that link to pirate sites. The day after the bill was introduced, Google put the following image on its search page for twenty-four hours. The image was viewed by 1.8 billion people.

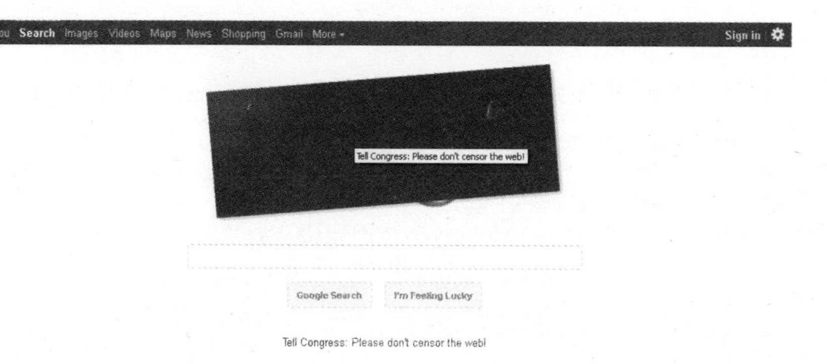

Note the use of the word *censor* and the "tell Congress" line, which enabled you to message your congressperson.

Needless to say, the email servers of Congress were overwhelmed, and on January 20, 2012, the chairman of the House Judiciary Committee, Lamar Smith, withdrew the bill. The very notion that getting Google to stop linking to criminal pirate sites constituted censorship is an exercise in Orwellian doublespeak. But the effect it had on legislators was to make them essentially Google captives. So much so, in fact, that Google was able to enlist many of them in its battle against the European Union, whose antitrust regulators seemed more willing to call Google a monopoly. As the *Guardian* reported, "Republican and Democratic senators and congressmen, many of whom have received significant campaign donations from Google totaling hundreds of thousands of dollars, leaned on parliament in a series of similar — and in some cases identical — letters sent to key members of the European Parliament."

2.

But it is in the area of "regulatory capture" that Google has potentially really excelled. Regulatory capture, according to Nobel laureate George Stigler, is the process by which regulatory agencies eventually come to be dominated by the very industries they were charged with regulating. Because of that dominance there is then a risk that the interests of those industries is, to an extent, protected. Putting aside the fact that Google chairman Eric Schmidt has visited the Obama White House more than any other corporate executive in America and that Google chief lobbyist Katherine Oyama was associate counsel to Vice President Joe Biden, the list of highly placed Googlers in the federal government is truly mind-boggling.

- The US chief technology officer and one of her deputies are former Google employees.
- The acting assistant attorney general in the Justice Department's antitrust division is a former antitrust attorney at Wilson Sonsini Goodrich & Rosati, the Silicon Valley firm that represented Google.
- The White House's chief digital officer is a former Google employee.
- One of the top assistants to the chairman of the FCC is a former Google employee and another ran a public lobbying firm funded in part by Google.
- The director of United States Digital Service, responsible for fixing and maintaining Healthcare.gov, is a former Google employee.
- The director of the US Patent and Trademark Office is the former head of patents at Google.

And of course the revolving door goes both in and out of the government, as the Google Transparency Project (an independent watchdog report) clearly stated.

- There have been fifty-three revolving-door moves between Google and the White House.
- Those moves involved twenty-two former White House officials who left the administration to work

for Google and thirty-one Google executives (or executives from Google's main outside firms) who joined the White House or were appointed to federal advisory boards.

- There have been twenty-eight revolving-door moves between Google and government that involve national security, intelligence, or the Department of Defense. Seven former national security and intelligence officials and eighteen Pentagon officials moved to Google, while three Google executives moved to the Defense Department.
- There have been twenty-three revolving-door moves between Google and the State Department during the Obama administration. Eighteen former State Department officials joined Google, while five Google officials took up senior posts at the State Department.
- There have been nine moves between either Google or its outside lobbying firms and the Federal Communications Commission, which handles a growing number of regulatory matters that have a major impact on the company's bottom line.

I am not suggesting any wrong-doing by an individual. It is simply a case of the regulators and the industry being dominated by like-minded people who see things the same way. Here one can sense that Google has a type of insurance policy: at key agencies such as the FCC, the Office of Management and Budget, the patent office, and the Justice Department's antitrust division, Google will always have, in effect, a seat at the table. And Google is casting its net beyond the Obama administration, according to an investigative report in *Quartz* that looked into Eric Schmidt's firm the Groundwork.

> The Groundwork, according to Democratic campaign operatives and technologists, is part of efforts by Schmidt—the executive chairman of Google parent-company Alphabet — to ensure that [Hillary] Clinton has the engineering talent needed to win the election. And it is one of a series of quiet investments by Schmidt that recognize how modern political campaigns are run, with data analytics and digital outreach as vital ingredients that allow candidates to find, court, and turn out critical voter blocs.

Google makes sure to place bets on both sides of the aisle. So while Eric Schmidt is advising Hillary Clinton's campaign, Larry Page flew with Sean Parker and Elon Musk in March of 2016 to a secret Republican meeting at a resort in Sea Island, Georgia, organized by the right-wing think tank the American Enterprise Institute. There they met with Republican leadership, including Mitch McConnell and Paul Ryan as well as Karl Rove, to plan Republican 2016 election strategy. My own experience in talking to legislators about Internet reform has led me to understand that Google, Amazon, and Facebook are deeply embedded in both parties, and their interests will be protected no matter who is in the White House.

But occasionally even these insurance policies are not enough to head off trouble. Since 2014, two examples of the way Google applies its political muscle have played out in public. The first was at the Federal Trade Commission, the agency that is supposed to regulate Google's advertising and search business. On March 24, 2015, the *Wall Street Journal* revealed the existence of a leaked report from the competition bureau of the FTC recommending that Google be prosecuted for abusing its market position by recommending Google services over those of third parties. What was astonishing was that the full commission had, in a very unusual manner, overruled the staff recommendation and decided against prosecuting Google. The *Journal* alleged that the 230 meetings that Google had had at the White House in the run-up to the complaint dismissal had influenced the commission. This caused the FTC to issue a classic denial.

> Today's Wall Street Journal article "Google Makes Most of Close Ties to White House" makes a number of misleading inferences and suggestions about the integrity of the FTC's investigation. The article suggests that a series of disparate and unrelated meetings involving FTC officials and executive branch officials or Google representatives somehow affected the Commission's decision to close the search investigation in early 2013.

But as most everyone – including the numerous travel services that competed against Google – knew, the FTC staff was right. As Yelp CEO Jeremy Stoppelman testified before the Senate antitrust subcommittee, "Google first began taking our content without permission a year ago. Despite public and private protests, Google gave the ultimatum that only a monopolist can give: In order to appear in web search you must allow us to use your content to compete against you." So Google did advantage its own services, and as it pushed into more markets, the problem would only increase. The final irony is that the FTC's decision not to prosecute Google was influenced heavily by a paper that Google paid Robert Bork (shortly before he died) and Gregory Sidak to write for them. Bork and Sidak's argument said: "[The fact] that consumers can switch to substitute search engines instantaneously and at zero cost constrains Google's ability and incentive to act anti-competitively." But as many researchers have pointed out, my use of multiple Google services (Gmail, Google Maps, and Google Calendar) essentially ties me to Google and makes the cost of switching far more cumbersome in terms of time and effort than Bork suggests.

A more blatant display of Google's power took place after Mississippi attorney general Jim Hood subpoenaed Google in October of 2014 in an effort to determine whether Google was in compliance with the Justice Department settlement the company had entered into in 2011 – i.e., whether it was still knowingly profiting from illegal trade. Google had paid a fine of $500 million, which covered the amount that it had profited from the advertising for illegal pharmacies on its search service, plus the amount that the Canadian pharmacies had received from US customers buying illegal drugs. All the oxycodone that was available online through a Google search a few years earlier had been stopped. So General Hood asked for, among other things, the search logs to determine whether Google was sending its customers to other illegal sites, including those selling pirated movies, music, and games.

Google, citing the DMCA, said it was protected under federal law and the First Amendment and that the subpoena was an attempt to coerce them into blocking sites that infringe on copyright. "The Attorney General may prefer a pre-filtered Internet," the lawsuit read, "but the Constitution and Congress have denied him the authority to mandate it." Hood quickly retreated but made it clear that Google was trying to use its money and power "to stop the State of Mississippi for daring to ask some questions." Nevertheless, he said he would call the company and try to work out a deal. Needless to say, when he called, Google didn't answer. There was no deal to be worked out, and after going back and forth in the court of appeals, both the subpoena and the lawsuit were dropped for good in 2016. As former labor secretary Robert Reich told me, "My sense is that political power trumps any ideology such as consumer welfare economics, but a powerful ideology can help sell the position advocated by a powerful political player. The irony here, of course, is that Google's entrenched position has given it enormous political power — which is one of the arguments against allowing any company to become so large and entrenched."

3.

If Google treats all content as a commodity against which it can place ads, what business is it really in? That's right:

ad sales. Because the real value of Google and Facebook lies in the data mining. For them the difference between the supreme artistry of a Martin Scorsese short film and an amateur cat video lies only in the number of views that can be sold to advertisers. So it is in the advertising business that Google has its second major advantage, after regulatory capture. The extent of Google's dominance in online advertising was laid bare on November 12, 2014. As the *Wall Street Journal* reported that day:

> A large swath of the Internet ran without advertising for over an hour Wednesday after Google's online ad-serving system DoubleClick for Publishers went down. The outage caused websites run by publishers including *BuzzFeed*, *Time* and *Forbes* to show blank spaces where display ads usually run... Wednesday's outages affected more than 55,000 websites, according to Dynatrace, which monitors website and web application performance for companies including eight out of the 10 largest retailers in North America.

Compared to its competitors, it is clear that Google owns the dominant share of online advertising. So instead of a competitive market for advertising keywords, Google runs an auction, which provides no transparency for the buyer other than the minimum price, set at Google's discretion.

But now Google is trying to extend its advertising monopoly into the world of TV. As I noted earlier, Google has key allies inside the FCC, which is now considering how to force cable and satellite companies to open up their set-top box business to outside competitors through a technology it calls AllVid. Google could essentially offer local restaurants, auto dealers, and other advertisers a targeted interactive ad on its TV Search page that would be far more efficient than advertising on local TV stations. The effect of this would be to totally kill the local TV spot-advertising market, which funds most local news programming. We have seen what these kinds of competitive services have done to the newspaper business. Without advertisements from local business to support the presence of reporters at city hall, much local news is no longer covered.

Consultant Scott Cleland, former deputy US coordinator for international communications and information policy in the first George Bush administration, sees the hand of Google in this proceeding:

> At its simplest, what the FCC AllVid proposal does is force the regulated pay TV industry to create a Google-most-friendly IP search interface/portal into their proprietary pay TV offerings. That way Google would be enabled to index and then monetize their competitors' most valuable proprietary information for free, by adding Google Internet ads as an overlay to competitors' ad-supported content, and by skipping the ads of their competitors, to kneecap their competition's relatively much smaller advertising businesses, and to devalue their competitors' paid content assets made vulnerable by the FCC proposal.

For a company that totally dominates the Internet ad business, the $73 billion TV ad market constitutes its only hope to continue the 20 percent annual growth rate that shareholders have come to expect.

In 2015, an article in the *Guardian* stated, "Were Google a manufacturer, say, a monopoly such as it has over internet search would never be allowed." But the ghost of Robert Bork is whispering in my ear: "Why does this matter to society? Where is the harm?" The same *Guardian* article says, "Google's dominance is self-reinforcing, but also makes it more useful. Larger audiences improve Google's data and make its products more accurate — as well as ever more impossible to avoid. As European competition commissioner Margrethe Vestager acknowledged last week, we live in the Google age." So the techno-determinist philosophy has even reached into the brain of Google's most determined regulatory opponent. This theory of self-reinforcing dominance has often been called the network effect, or Metcalfe's law, after PARC researcher Bob Metcalfe's formula that the value of a network is proportional to the square of the number of users. As more people use Google's search engine, the company becomes exponentially more valuable. Could it be that Google is now what economists call a natural monopoly — a firm, such as a utility, that can supply an entire market's demand for a service at a price lower than two firms could? In general, utilities are regulated by the government to protect consumers.

As a society we are going to have to decide fairly soon whether Google, Facebook, and Amazon form a natural monopoly that needs to be regulated or whether we are going to pretend that competition and capitalism can exist in harmony in the digital age. They are existing in harmony today because the counternarrative about the costs of digital innovation — and who bears those costs — has not been made. Peter Thiel knows that competition is anathema to the kind of capitalism he wants, but the regulators in Washington still live in a fantasy world of "perfect markets." Gary Reback, one of the most prominent antitrust attorneys in Silicon Valley, has argued that they don't. Reback noted to the *New York Times,* "Once one of these companies gets a monopoly, it's easy to spread the monopoly to adjacent markets by acquisition. You would think antitrust enforcers would know this by now." Do the regulators not see Google's monopoly as a problem, or do they believe that Google's political power is so great that the company is untouchable? Digital monopolies can either be part of the problem or part of the solution. But I'm doubtful that we can rely on their declarations of "Don't be evil" while continuing this regime of unregulated monopoly. At this point, the economy in which we all live is shaped along the contours of Google, Amazon, and Facebook dominance. To propose that these firms trim their sails is to risk introducing chaos to an economy that not too long ago was brought to its knees.

4.

In the summer of 1996 I founded with two friends (Richard Baskin and Jeremiah Chechik) one of the first streaming-video-on-demand companies, Intertainer. We raised a good bit of money from Comcast, Sony, Intel,

NBC, and Microsoft and had a working service to thousands of broadband customers in the fall of 1998. Unlike Napster or YouTube, we believed in asking permission, so we licensed movies and television programs from most of the major studios and paid them large guarantees for the service. I started the company because I had seen a demonstration at CableLab of video streaming over an early broadband test bed. The quality of the video wasn't perfect, but the idea that you could get whatever video you wanted on demand, instead of going down to your local Blockbuster store, seemed pretty compelling. The people who built Intertainer were an extraordinary group of pioneers. They invented interactive tools that today form the basis for the video ad systems on the Internet. And pretty soon we were doing something – sending high-quality video over the Internet – that had never been done before. An article about Intertainer in *Business 2.0* in December of 2001 seems almost ironic when read from our present vantage point.

> For sheer cockiness, the business plan that Jonathan Taplin hatched with two friends in the summer of 1996 ranked right up there. Intertainer, as they named their company, promised nothing less than instant access to a vast electronic storehouse of movies, documentaries, and other fare, right from your own home. A couple of clicks and, to hear Taplin tell it, you could be watching Almost Famous or Ken Burns's jazz series or the latest Eric Clapton video or the final game of the 1975 World Series, all with the VCR-like ability to interrupt the show whenever you felt like going out for a beer.

> The basic idea, called video on demand, or VOD, had been knocking around for years, as had Taplin himself. He was in his mid-40s — almost his dotage, by the entrepreneurial standards of the Internet Age — and he had suspiciously close ties to the entertainment industry establishment. He conceded nothing to his juniors, however, when it came to evangelical fervor. And indeed, if you stop to think about what home entertainment is and what VOD could be, the implications are dizzying. Such things as channels and schedules could eventually wither away. The whole idea of "watching TV" — that is, turning on the set and seeing what's "on" — could come to be regarded as a peculiar and slovenly habit of the medium's early days.

But of course, as any pioneer knows, making one's vision a reality is never easy. In early 2002, when we had around 150,000 customers, we heard a rumor that one of our shareholders, Sony, was quietly developing a service that looked just like Intertainer. When I asked them about it, they pleaded ignorance. And then one day in the summer of 2002 they announced the service, called Movielink, in a joint venture with Warner Bros., Universal Studios, MGM, and Paramount Pictures. Soon after that, the major studios stopped licensing films to Intertainer, and the service was forced to close down. This broke my heart, and I refused to take it lying down.

In September of 2002, Intertainer filed an antitrust suit against three major entertainment companies, AOL Time Warner, Vivendi Universal, and Sony as well as their wholly owned service Movielink in the United States District

Court for the Central District of California, Western Division, accusing them of conspiracy to fix prices in the digital distribution of entertainment and restraint of trade. In March of 2006, the defendants reached an out-of-court settlement with Intertainer. All I am allowed to say about the settlement is that "the lawsuit was resolved to the satisfaction of all parties." It certainly was to the satisfaction of Intertainer, which is why I believe that well-enforced antitrust law can really help the little guy compete against corporate giants. Intertainer still exists and is currently in the business of licensing its video-on-demand patent portfolio. Companies such as Microsoft, Comcast, Thomson Reuters, Viacom, Discovery Communications, and Apple are currently license holders.

After years as a music and film producer, with a detour in the corporate world, my goal was to create a nexus between technology and entertainment. I knew that the combination of the digitization of all the world's film libraries combined with a Moore's law–type improvement in the quality of video streamed over the Internet would lead to a much more efficient video distribution system. Like my investors, I knew that creating a platform for the consumption of film was the place to innovate. We did that, and in the process we helped radically improve both the video quality and the back-end systems that allow users to get (as our first motto stated rather boldly) "anything, anytime, anywhere."

Ultimately, as one executive from a company that licensed films to us put it, "the industry does not want an online HBO – another middleman." Larger players in the industry, working in concert with one of our own investors, formed a cartel that made it impossible for us to do business. Monopoly squelches innovation in all industries, and I saw just that happen at an early moment in the digital migration of entertainment. Perhaps if we had been doing business in 2010, we would have been bought rather than forced to close down. The counterfactuals for our current monopolistic environment are hard to accurately envision, but I lived one.

CHAPTER EIGHT

The Social Media Revolution

The social norm (of privacy) has evolved over time.
— Mark Zuckerberg

1.

Yael Maguire, director of Facebook's Connectivity Lab, is making a presentation to a *Fast Company* reporter. "There is only 10 percent of the world that is not able to connect if they pulled out a phone. Our job is to figure out how to connect the last 10 percent." His solution: the prototype of the Aquila, an ultra-lightweight carbon-fiber drone with a 138-foot wingspan (a Boeing 737 has a 113-foot wingspan) that weighs only 880 pounds. With the right battery technology, the drone should be able to hover for three months over a remote village in India and provide a basic Internet service Facebook calls Free Basics. No one ever said Facebook CEO Mark Zuckerberg lacked ambition.

But now citizens of that remote village are about to

enter the surveillance society that billions of us have been inhabiting since 2000. "Who could possibly be against this?" Zuckerberg asked in an editorial in the *Times of India* on December 28, 2015. As it turns out, many in India were opposed. According to an article in *The Economist,* "Critics of the program say that Facebook's generosity is cover for a land grab. They argue that Free Basics is a walled garden of Facebook-approved content, that it breaches consumer privacy by sucking up all the data generated by users of the service, and that it is anticompetitive to boot." When the Indian government finally ruled that Facebook should not be able to "shape users' Internet experience" by providing only a limited set of sites, Facebook board member Marc Andreessen was outraged, tweeting, "Anti-colonialism has been economically catastrophic for the Indian people for decades. Why stop now?" But even Andreessen was shocked by the backlash to his tweet, much of it from the Indian tech community in Silicon Valley. He unwittingly revealed a previously unspoken truth: Facebook and Google are the new colonial powers.

The battle for world domination is on. If Google's Larry Page worries about any competitor, it is probably Mark Zuckerberg. While Facebook began as a way to make social networks visible and to ease communication between them, it is now — like Google — in the surveillance marketing business. Facebook and Google sell the data you give them to marketers. Google gets the data through your search history. Facebook gets it through your social media posts. The scale of the Facebook ecosystem — which includes WhatsApp, Messenger, and Instagram — is astonishing: 1.6 billion users of Facebook itself; 1 billion on WhatsApp; 900 million on Messenger; 400 million on Instagram. Facebook controls more than 75 percent of US mobile social media platforms. Under any normal antitrust regime this would be considered a monopoly. Like Google, Facebook has taken to presenting itself as a public service. "Don't be evil." "Who could be against it?" But a 2014 survey by online identity manager MyLife shows that 82.9 percent of those polled said that they did not trust Facebook with their personal data. I am a Facebook user, and in many ways I think it is a wonderful tool for communication. I also suspect you will find that Mark Zuckerberg, the brash young man who founded the company at the age of twenty, is growing up and becoming aware of the awesome responsibilities he has in running the world's largest social network.

The bratty kid portrayed in the movie *The Social Network* may have been changed by marriage and fatherhood. Larry Page, Peter Thiel, and Jeff Bezos are in their forties and fifties. Their libertarian ideals are fairly fixed, but watching the evolution of Mark Zuckerberg over the past ten years is to see a maturation process in both the man and his company. I may be a fool — perhaps he doesn't really believe that his job is to bring the twenty-first century to the four billion people who remain offline. But even if I'm wrong, and this is all some amazing publicity stunt, the process of connecting two billion people on one site will force Zuckerberg to deal with three critical elements of our story: the role of privacy in our lives, the role advertising will play in the media landscape, and the future of communications.

2.

The curly-haired Harvard sophomore Mark Zuckerberg, who in the fall of 2003 wrote the program for a site called Facemash, would hardly have been picked out as a future *Time* magazine Person of the Year. With green eyes in a sea of freckles, his head seems a bit big for his slender frame. Most of his life he has dressed in a uniform of gray T-shirt, baggy jeans, and Adidas flip-flops. He conceived Facemash in a fit of pique over a girl who had put him down and programmed the website to compare Harvard students for "hotness." Using code that had been designed to compare computer chess players, Facemash would show you two faces of the same sex and allow you to rate which was hotter. Then the one you picked would be matched with a new opponent. In the first iteration of the program, an occasional farm animal would appear. In classic "don't ask permission" form, Zuckerberg ripped all the photos off hacked online "facebooks" maintained by each of the Harvard residential houses (dorms). These were classic bad driver's license–style pictures taken at orientation. After deleting the farm animals, Zuckerberg launched the site, running it off his laptop from his dorm room over the Harvard network.

In the first several hours the site was up, it was visited by 450 students who had voted on twenty-two thousand pairs of photographs. The computer services department couldn't figure out why the school's network was slowing down until they traced it to Zuckerberg's dorm room, at which point they shut off his Web access. The next morning, after complaints of sexism, racism, and general stupidity, Zuckerberg was on the cover of the *Harvard Crimson* newspaper and the subject of an editorial saying that he was "catering to the worst side of Harvard students." It was perhaps one of the shortest-lived websites in history, but we need to understand that beyond the adolescent-prank nature of it, Zuckerberg might have understood the basic narcissism of young people better than the editors of the *Harvard Crimson* did. After being put on probation by Harvard, he publicly apologized to women's groups and agreed to see a counselor. He celebrated his light sentence with his roommates and a bottle of Dom Perignon. The excitement of making something that was immediately popular was in his brain. He wanted more.

Within a couple of months, he was building Thefacebook, the initial Harvard-only version of the platform that now has 1.5 billion users. Thefacebook was not the first social network – Friendster already had more than 3 million users, and Myspace was just getting started. But Zuckerberg understood three things that helped him succeed where the first movers would fail. First was simplicity of design. The clean look of Thefacebook contrasted with the cluttered, almost anarchic design of Myspace. What Myspace thought was a positive feature (the ability for anyone to elaborately design his or her own page) turned out to be a bug. The simplicity also made the server load less problematic. The pages loaded quickly, whereas Friendster often took a minute or more to load. Zuckerberg knew the average multitasking college student was borderline ADHD, so speed was critical.

The second thing Zuckerberg understood was that by releasing the service only on elite college campuses (he launched in most of the Ivy League schools within two months), he could not only appeal to the snobbism of students but also take advantage of the dense social networks that are native to college campuses. The desire to know what your friends are doing on a Thursday night is probably more intense on a college campus than it is at any other location. Also, by controlling the rollout, he was able to avoid the technical crashes that had hurt Friendster. It allowed Zuckerberg and his roommates (who had joined the company) to achieve critical mass without spending millions of dollars on server capacity.

Finally, Thefacebook team grasped the inherent utility that the site would serve. In a piece published in the *Harvard Crimson* only two weeks after the site went live, Amelia Lester wrote, "There's little wonder why Harvard students, in particular, find the opportunity to fashion an online persona such a tantalizing prospect.... [Thefacebook is] about performing and letting the world know we're important individuals. In short, it's what Harvard students do best." But it turned out it wasn't just elite Harvard kids who wanted to fashion an online persona – it was everyone.

When Thefacebook really started to grow, in the late spring of 2004, Zuckerberg and his right-hand man, Dustin Moskovitz, decided to go to Silicon Valley for the summer. Zuckerberg had met Sean Parker in a Chinese restaurant in New York in May and had been awed by his outlaw tales of

Napster. Zuckerberg had written a music-recommendation engine while he was a senior at Exeter, and so Napster loomed large in his notion of hipness. When the two men got to Palo Alto in June, they ran into Parker, who was essentially homeless, having been thrown out of his latest company, Plaxo, an online address-book application. It is a tribute to Zuckerberg's naive trust that he invited Parker to live in the house he and Moskovitz had rented. Parker promised to teach them about the shark tank known as Sand Hill Road — the center of the Valley's venture capital business. In that role he did two important things. First, he kept Zuckerberg centered on Facebook, even though the young coder was spending a lot of time writing another program called Wirehog, which was essentially a version of Napster. Parker convinced him that it would be nothing but trouble in the form of lawsuits from the content community. The genius of Facebook was that the users provided all the content. There was no need to steal pictures, as he had for Facemash, or music files, as Parker had done for Napster. The second thing Parker did was to introduce Zuckerberg to Peter Thiel.

Peter Thiel grasped almost immediately the potential of Facebook. As David Kirkpatrick explained in *The Facebook Effect: The Inside Story of the Company That Is Connecting the World*, "What happened once it opened at a new school was what most impressed Thiel. Within days it typically captured essentially the entire student body, and more than 80 percent of users returned to the site daily!" This was unprecedented, and Thiel knew that Facebook fit all four of his success categories. It had proprietary technology with network effects that could scale, and it had a good brand. Sean Parker then solidified the brand identity by getting Zuckerberg to drop "The" from the name. It then became just Facebook. Thiel immediately gave the company a loan of $500,000, which was convertible into 10 percent of the stock. He did it as a loan because Zuckerberg had a lot of messes to clean up with Harvard students who claimed that he had stolen their idea.

3.

Before I continue with Zuckerberg's story, I want to pause to ponder just why his invention created such a profound shift in the world of communication. Facebook is changing the norms of what it means to be private, what it means to be a kid, and what it means to be a "human product." As the phrase goes, if you are not paying for it, you are not the customer, you are the product. Perhaps Mark Zuckerberg's greatest insight was that the human desire to be "liked" was so strong that Facebook's users would create all the content on the site for free. In 2014, Facebook's 1.23 billion regular users logged in to the site for seventeen minutes each day — as the *New York Times* pointed out. In total, that's more than *39,757 years* of time collectively spent on Facebook in a single day. That's almost fifteen million years of free labor per year. Karl Marx would have been totally mystified.

The first question that needs to be asked is why we are

so willing to surrender our labor and our personal data for free to an immensely profitable monopoly. To understand that, researchers at the University of Connecticut, led by Daniel Hunt, tried to understand why people spend so much time on Facebook. The blog *ReadWrite* summarized the study like so:

> Researchers have long known that five broad categories drive online activity: information seeking, interpersonal communication, self-expression, passing time and entertainment. In the study led by Hunt, the goal was to see if the same measures drove people to spend time on Facebook. The study confirmed that, with the exception of information seeking, all of the other behavioral factors that drive online activity hold true for Facebook, with entertainment and time passing being two of the biggest drivers of Facebook activity.

What their research really found was that although we may originally sign up for Facebook for interpersonal communication, fairly quickly we use it to fight boredom. But fighting boredom could not be the only reason, because young adults have many entertainment options. What distinguishes Facebook and other social networks is their role in self-expression — the need to present oneself to one's peers in a positive light. In 1987, the psychologists Hazel Markus and Paula Nurius suggested that a person has two selves: the "now self" and the "possible self." What Facebook allows is the presentation of the "hoped-for possible self" in the form of the best-looking selfies, the coolest party pictures, and other presentations of an ideal life that may or may not exist.

But the presentation of self may not always be voluntary. In her book *Dragnet Nation: A Quest for Privacy, Security, and Freedom in a World of Relentless Surveillance,* Julia Angwin presents the story of Bobbi Duncan:

> Bobbi Duncan, a twenty-two-year-old lesbian student at the University of Texas, Austin, tried to keep her sexual orientation secret from her family. But Facebook inadvertently outed her when the president of the Queer Chorus on campus added her to the choir's Facebook group. Bobbi didn't know that a friend could add her to a group without her approval and that Facebook would then send a note to her entire list of friends — including her father — announcing that she had joined.

Her father's note on his Facebook page — "Hell awaits you pervert, good luck singing there" — makes it clear why Bobbi wanted to keep her sexual orientation to herself. Part of the reason she was unable to do so is because Zuckerberg and his Facebook team had embraced the notion of "radical transparency" — the idea that openness is the chief goal of the service and thus you can only use your real name. But many people, including Bobbi Duncan, disagree. Zuckerberg told Kirkpatrick, "To get to this point where there is more openness — that's a big challenge, but I think we will do it. The concept that the world will be better if you share more is something that is pretty foreign to a lot of people and it runs into all these privacy concerns."

The disdain for "these privacy concerns" first surfaced when in 2007 Facebook deployed an application called Beacon. This was essentially an alert system that told your friends you had purchased something on a partner site. It was built as an opt-out system, so you actively had to tell Facebook each time you didn't want the site to broadcast your purchase to all your friends. It was a total disaster from the outset, but Zuckerberg was so confident that he knew better than his users that he refused to turn it off for many weeks while the PR disaster escalated. Eventually he relented and posted a mea culpa on his blog, saying, "We've made a lot of mistakes building this feature, but we've made even more with how we've handled them." Despite Zuckerberg's regret and a payment of $9.5 million in a class-action suit over Beacon, many who worked with him feel he doesn't really understand privacy. Charlie Cheever, one of his key programmers, told Kirkpatrick, "I feel that Mark doesn't believe in privacy that much, or at least believes in privacy as a stepping-stone [to radical transparency]."

The privacy issue was reignited in early 2014, when the *Wall Street Journal* reported that Facebook had conducted a massive social-science experiment on nearly seven hundred thousand of its users.

> To determine whether it could alter the emotional state of its users and prompt them to post either more positive or

> negative content, the site's data scientists enabled an algorithm, for one week, to automatically omit content that contained words associated with either positive or negative emotions from the central news feeds of 689,003 users.

As it turned out, the experiment was very "successful" in that it was relatively easy to manipulate users' emotions, but the backlash from the blogosphere was horrendous. "Apparently what many of us feared is already a reality: Facebook is using us as lab rats, and not just to figure out which ads we'll respond to but to actually change our emotions," wrote Sophie Weiner on AnimalNewYork.com.

In May of 2016, Facebook board member Peter Thiel was drawn into the privacy debate when it was revealed that he had financed a lawsuit brought by Hulk Hogan against the online news site *Gawker,* which he accused of violating his privacy when it put a sex tape online. Thiel's stated reason was to destroy *Gawker,* one of whose publications, *Valleywag,* had outed Thiel in 2007. After that incident Thiel had stated, "*Valleywag* is the Silicon Valley equivalent of al-Qaida." Nine years later he got his revenge, telling the *New York Times,* "I saw Gawker pioneer a unique and incredibly damaging way of getting attention by bullying people even when there was no connection with the public interest." The revelation quickly exposed the battle lines between Silicon Valley and the media. Mark Zuckerberg had publicly stated that privacy norms were evolving and his board member Thiel had backed him up. But with *Gawker,* Thiel was reasserting his right to privacy, even though it was general knowledge in Silicon Valley that he was gay. The irony was that Owen Thomas, the gay writer who wrote the *Valleywag* piece, ended with this statement: "That's why I think it's important to say this: Peter Thiel, the smartest VC in the world, is gay. More power to him."

After Thiel confessed to secretly funding the *Gawker* lawsuit, Marc Andreessen, Thiel's fellow Facebook board member, came to his defense, writing on Twitter, "Greenpeace and Sierra Club, among many other progressive groups, routinely fund other plaintiffs' lawsuits." But Andreessen neglected to mention that they do so openly, not in secret. Another VC, Vinod Khosla, tweeted support, remarking, "The press gets very uppity when challenged." But it was the far-right supporters of Donald Trump (Thiel was a Trump delegate) who came to Thiel's support in great numbers on Twitter, using the hashtag #ThankYouPeter. Trump's principal Breitbart News Network backer, Milo Yiannopoulos, wrote, "Freedom-minded PayPal founder Peter Thiel has revealed himself to be a Batman of sorts. The hero Silicon Valley needed." But Jason Mandell, a PR specialist based in Silicon Valley, highlighted the basic contradiction between libertarian beliefs and freedom of the press. "People like Peter Thiel are used to being able to tell an engineer 'this is broken – fix it'," Mandell said. "They don't understand the unique dynamic between the press and the public. They don't understand the first amendment and free speech as it relates to the media."

Of course the very fact that Thiel's candidate Donald Trump has threatened to remove libel protection for media organizations should give one pause. In February of 2016, at a rally Trump said, "One of the things I'm going to do if I win... I'm going to open up our libel laws so when they write purposely negative and horrible and false articles, we can sue them and win lots of money." Nicholas Lemann, writing in the *New Yorker* in May of 2016, understood how high the stakes were: "Remember that Thiel is a graduate of Stanford Law School who clerked for a year on the Eleventh Circuit, and that, in his world, 'scale' and 'disruption' are the hoped-for ends of every investment. He is surely aware of this case's potential to begin a reexamination of the fundamental questions in American press law, far beyond the fate of Gawker."

The current narrative we seem to tell ourselves about our privacy is that it is a sort of currency we trade to corporations in return for innovation. But the corporation has an insatiable appetite for our most personal data in order to drive us to consume during our every waking moment. I think this is critical, because in some ways social networks are powerful engines of conformity. The ability for students to develop their own ideas, identities, and political affiliations should take place outside of the panopticon view of Facebook, but whether this is any longer possible is an open question. My own memory is that the development of my political and cultural persona between the ages of fifteen and twenty-one had a lot to do with being outside the zone of judgment of my parents, their conservative peers from my hometown, Cleveland, and maybe even from my siblings. I'm not sure that it could happen if we were all on Facebook together.

4.

Now, it very well may be that privacy is a hopelessly outdated notion and that Mark Zuckerberg's belief that privacy is no longer a social norm has won the day. But it is one thing to forfeit our privacy as individuals to a company that we believe is delivering a needed service and another to open our personal lives to the federal government.

Has Zuckerberg's belief in the end of privacy influenced his relations with government surveillance? In June of 2013, Glenn Greenwald, writing in the *Guardian*, revealed that in 2009, Facebook, along with Google and Apple (and four other online service providers), had given the National Security Agency direct access to their worldwide network for the agency's PRISM spying program. Below is a slide from an NSA internal presentation.

Writing in the British publication *Computer Weekly,*

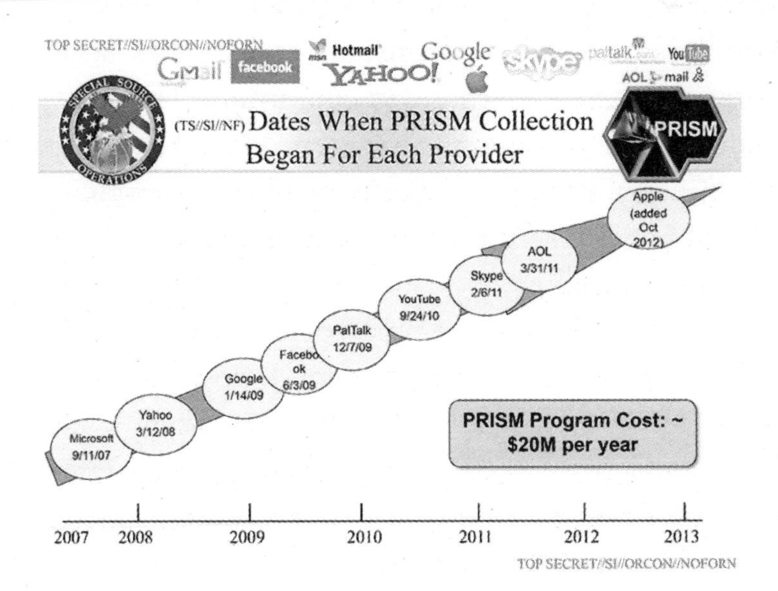

Kevin Cahill noted, "The possibility of universal mass surveillance became clear to the NSA in the early part of this century, as the reach of the nine largest US Internet providers grew to more than 50% of all Internet users on the planet. The Internet giants were developing a reach that the spooks in Fort Meade, the NSA's HQ in Maryland, could only dream of. So the spooks stepped in and more than nine corporations obliged by delivering their clients' data. It was, and is, the largest hack in human history – so far."

Where does surveillance marketing stop and spying begin? When Greenwald's article, based on the revelations of Edward Snowden, came out, Zuckerberg responded that the government had done a "bad job" of protecting people's privacy. "Frankly I think the government blew it," he said. But just how hard did he fight to protect the data of his users? And had Facebook, in its commercial hoovering of consumer data, done an equally bad job of protecting its users' privacy? This also seems to be why Apple didn't receive more vocal support from Facebook and Google in February of 2016, when it chose to resist the FBI's request that it build a trapdoor into the encryption of the iPhone 6. As one wag posted on Twitter, "Why haven't Ooogle, 2Facedbook and Microsuck got [Apple CEO] Cook's back? Their biz model: spying & selling data! Appl sells phones!"

5.

To answer the question of why Facebook is "spying and selling data," one must understand why they need all your data in the first place. The answer is simple. Facebook's business model relies on selling advertising at a higher rate than most comparable Internet sites. It is here that the basic competition between Facebook and Google plays itself out. Whichever company can accumulate the most data on you can then sell highly pinpointed advertising at the highest price. For the moment, Facebook seems to be winning. At a time when Google ad rates are falling, Facebook announced in the fourth quarter of 2015 that "the average price per ad increased 21%, while total ad impressions increased 29% on a year over year basis." The reason they get this premium is by microtargeting you. If I want to reach women between the ages of twenty-five and thirty in zip code 37206 who like country music and drink bourbon, Facebook can do that. Moreover, Facebook can often get friends of these women to post a "sponsored story" on a targeted consumer's news feed, so it doesn't feel like an ad. As Zuckerberg said when he introduced Facebook Ads, "Nothing influences people more than a recommendation from a trusted friend. A trusted referral is the Holy Grail of advertising." Essentially Facebook has built a massive database of the consumer preferences of two billion people. As Bob Garfield, one of the most respected analysts in the advertising business, explains:

> This theoretically should be good for consumer marketers, and arguably helpful for consumers, notwithstanding the creepy sense of being shadowed through life. It also gives government or sinister data thieves a lot of data to get hold of and abuse. The intersection of surveillance

> marketing, the security state and offshore organized crime suggests a scenario that Orwell, Koestler, Kafka, Huxley and Solzhenitsyn couldn't have imagined at a peyote party.

Bob, who cohosts *On the Media* for NPR and was a columnist for *Advertising Age,* has never been known to mince words, which is why he is so helpful here. When the flacks at Google and Facebook defend their privacy policies, they always note that all the data is "anonymized." But Garfield's amusing response to that is right on:

> At the current trajectory, barring legislative, regulatory or judicial intervention, databases will grow exponentially and their use will become precise. The data profiles will remain, ahem, nominally anonymous, because no marketer has any interest in knowing the name of the consumer being monitored and targeted — merely that person's checking account. That said, by overlaying databases of online activity, geolocation, purchasing behavior, etc., they will know more about your IP address than your wife knows about you.

The problem of targeted advertising on the Internet extends far beyond Facebook and Google, though they are the two dominant players. In the first quarter of 2016, eighty-five cents of every new dollar spent on online advertising went to Google or Facebook, according to Brian Nowak, a Morgan Stanley analyst. So all providers of content, be they musicians, filmmakers, journalists, or photographers, have to deal with Google or Facebook if they want to attract an audience.

For providers of content, programmatic advertising opens up a new battlefield in the fight to capitalize on your data. Needless to say you probably wonder why you are being shadowed by certain advertisers, no matter what website you are on. Let's say you're thinking of going to Las Vegas, so you check out prices at a Vegas hotel but don't make a purchase. You now have a Vegas cookie on your computer. Next time you log on, as an ad loads in your Web browser, information about the page it is on and your cookies are passed to an ad exchange, which auctions that information off in real time to the Vegas advertiser willing to pay the highest price for it. The winning bidder's ad is then loaded into the Web page nearly instantly. All this takes a few milliseconds. This is programmatic advertising, and there are two basic problems with the system, which now dominates Internet advertising.

First, it disadvantages quality content, because the advertiser is not interested in the content of the site. It is solely interested in targeting you. So there is no differentiation between NYTimes.com and a porn site. The *New York Times* spends millions of dollars on its content and expects to receive premium ad rates based on the quality "environment" those ads will be featured in. But programmatic advertising destroys that whole value proposition.

The second problem for advertisers and content owners is the fraud issue, which almost no one in the advertising business wants to talk about. A 2015 article in *Bloomberg Businessweek* nailed the problem:

> Increasingly, digital ad viewers aren't human. A study done last year in conjunction with the Association of National Advertisers embedded billions of digital ads with code designed to determine who or what was seeing them. Eleven percent of display ads and almost a quarter of video ads were "viewed" by software, not people. According to the ANA study, which was conducted by the security firm White Ops and is titled The Bot Baseline: Fraud In Digital Advertising, fake traffic will cost advertisers $6.3 billion this year.

How could it be that American brands are willing to spend $6.3 billion advertising to bots? The first reason is that the current "ad tech" business is like the Wild West during the Gold Rush – filled with scammers. But all this is going to change. At a conference in February of 2015, LUMA Partners CEO Terry Kawaja noted that out of more than two thousand ad tech companies in the market, just 150 might survive. *Digiday* reported his speech with the ominous headline WINTER IS COMING FOR THE AD TECH INDUSTRY. The beneficiary of the consolidation will be Google and its ad tech subsidiary, DoubleClick, which together already have a huge share of this business and will be able to buy the smaller players when "winter" comes. Here is Bob Garfield again:

> I think legislators, regulators and litigators will eventually (and maybe soon) begin to bring the hammer down. Furthermore, as the world moves to mobile, which is a more cookie-free environment, the tracking is more difficult. I

> think ad tech is its own worst enemy. Bot fraud, programmatic buying and ad blocking are costing marketers billions and compromising any advertising's ability to reach actual humans, which means that the very digital marketing system that represents the threat is itself on a bad trajectory, perhaps toward implosion.

Garfield's caution about ad blockers seems prescient. The digital marketing consultant Tune is projecting that ad blocking could reach 80 percent of smartphone owners by the third quarter of 2017. Apple has positioned itself in opposition to Google and Facebook in the ad-blocking wars because they have almost no advertising income. They support many ad-blocking apps on both the iPhone and iPad, much to the consternation of Google and the programmatic ad business. If the world of programmatic advertising does implode, it could look pretty ugly. But for publishers of quality content such as the *New York Times,* it could be a boon. The notion that advertising rates should be higher in an environment of quality has for years helped the publishers of the *Times* as well as companies such as Condé Nast. The world of programmatic advertising destroyed that connection, so maybe its demise might restore the premium rates that should accrue to high-quality content. But even if that were true, the best publications increasingly rely on Facebook to get readers. In 2014 online news sources such as *BuzzFeed* and *Huffington Post* got almost 50 percent of their inbound traffic from Facebook, and as the CEO of Bloomberg Media Group, Justin Smith, said, "The list is a lot longer than is publicly known of those that have Facebook delivering half to two-thirds of their traffic right now."

6.

So will Facebook really determine the future of journalism? What seems obvious in a world of *BuzzFeed* and *Huffington Post* being fed by Facebook is that the winning strategy seems to be to produce more content at a lower price. *Digiday* looked at the race for what some are calling peak content. What it found was that in 2010 the *New York Times,* with 1,100 people employed in the newsroom, created 350 pieces of original content per day and attracted 17.4 million page views per day. By contrast, the *Huffington Post,* with 532 people employed, posted 1,200 pieces of content per day (most of it created by third-party sites) and 400 blog entries (mostly unpaid), receiving 43.4 million page views per day. One can understand why the future of original journalism is threatened.

All this publishing may be attracting an audience, but as Steve Rubel, chief content strategist for Edelman, points out, "It's our view that increasingly, content publishing is only visible if it makes it out to the platforms where most of the time, attention and dollars are now going." And the two controlling platforms are Facebook and Google, although Apple is making an effort to move into this space. Visibility thus leads to profitability. This again raises the question I asked earlier: Will Facebook become a "rentier," charging publishers for access to its 1.6 billion users? But maybe the larger question is: Is this emphasis on quantity of output shoved onto platforms such as Facebook making us more or less informed? Does click bait coarsen our culture or just provide more amusement to alleviate our seemingly endless boredom? A tyrannical editor, equipped with all this data from Facebook about which article got the most "likes," can turn a newsroom into a sweatshop where piece count is the measure of your productivity and determines your salary. This seems to me to be a rather dispiriting vision of our media future and one that Mark Zuckerberg, who reigns over the funnel through which almost every publishing enterprise pours, must reckon with. Evan Williams, one of the founders of Twitter, now running *Medium,* told the *Guardian* that he is worried about these "feedback loops":

> If you look at feedback loops like likes and retweets, they've been very carefully crafted to maximize certain types of behaviors. But if we reward people based on a measurement system where there's literally no difference between a one-second page view or reading something that brought them value or changed their mind, it's like – your job is feeding people, but all you're measuring is maximizing calorie delivery. So what you'd learn is that junk food is more efficient than healthy, nourishing food.

If we are at peak content – the point at which the glut of things to read, watch, and listen to becomes completely unsustainable – then Zuckerberg will have to rethink his model. In August of 2016 Facebook announced it was changing its news-feed algorithm to try to cut down on the amount of click bait that appears on the site. It remains to be seen how this will affect quality journalism organizations that are dependent on Facebook traffic.

For the time being, Zuckerberg's answer seems to be to take even more control over news content. In March of 2015 Facebook partnered with news organizations such as the *New York Times, National Geographic,* and *BuzzFeed* to host content inside Facebook rather than linking to the news companies' sites. Ostensibly the reason was that this way stories would load faster on a mobile device, and many news organizations found this to be an attractive proposition. Especially for a site like *BuzzFeed,* which is not getting a lot of advertising on its own site but rather creates custom stories that are essentially advertising disguised as editorial content (called native advertising in the trade), one can see the advantage of embedding the stories inside Facebook. But for real news organizations like the *New York Times,* the bargain is fraught with peril. Will Oremus, writing in *Slate,* explains the dilemma.

> News sites aren't blind. They know it would be bad for them to cede control of their content — and, by extension, their relationships with readers and advertisers — to Facebook. And so if they could all get together and decide, as a group, what to do about Facebook, no doubt they'd think long and hard about the long-term sacrifices...But news sites don't function as a collective. On the contrary, they operate as rivals for the same audiences and advertisers. And Facebook has made it clear that those who sign on early will see huge growth in their Facebook reach. If that

> proves true, others will scramble to follow, even as it becomes clear they're seeing diminishing returns. Meanwhile, the holdouts would see their Facebook audiences wither and die, as Facebook's algorithms gradually downgrade posts that link out to third-party websites. Ultimately, links may become all but obsolete in the Facebook news feed.

Here we see the power of monopoly played out in the clear. What alternative does the *Times* have when faced with this choice? Emily Bell of the Columbia Journalism School, in an article entitled "Facebook Is Eating the World," wrote, "I can imagine we will see news companies totally abandoning production capacity, technology capacity, and even advertising departments, and delegating it all to third-party platforms [such as Facebook] in an attempt to stay afloat." I am hoping that the young CEO of Facebook will be willing to pause and think about where his company is taking the media business. Clearly most publications have come to accept that Facebook is a critical part of their audience-traffic ecology, but if they abandon attempts to build an audience for their own sites and just become article feeders for Facebook, they will eventually find themselves without a business rationale. As the University of California economist J. Bradford DeLong explained to me in a series of tweets, "Capitalism for information goods often goes badly wrong."

> It's a standard saying in Silicon Valley: If you aren't paying, you aren't the customer, you're the product. In a capitalist

> economy, sellers have strong incentives to fulfill the needs of their customers — they want them back. But here the customers aren't the viewers. The customers are the advertisers, who care whether the viewers watch, not whether the viewers are well informed.

This is a problem for me because for the past twelve years I have been a professor at the Annenberg School for Communication and Journalism at the University of Southern California. We are training a new generation of journalists for future jobs at places such as the *Los Angeles Times* and CNN — jobs that may not exist by the time they graduate. The Bureau of Labor Statistics has said that newsroom employment at the nation's 1,375 dailies could fall below 28,000, less than half the number of its high point in 1990. We are not teaching them how to write snarky tweets to gain followers but rather how to conduct an interview, write a lede, and shoot a short video. As Leon Wieseltier once wrote, "As the frequency of expression grows, the force of expression diminishes." And as President Obama noted in a speech, "Ten, twenty, fifty years from now, no one seeking to understand our age is going to be searching the tweets that got the most retweets." We certainly are not preparing our students for a career writing "sponsored content," like the kind Jacob Silverman described in *The Baffler*:

> Also called native advertising, sponsored content borrows the look, the name recognition, and even the staff of its host publication to push brand messages on unsuspecting viewers. Forget old-fashioned banner ads, those most reviled of

> early Internet artifacts. This is vertically integrated, barely disclaimed content marketing, and it's here to solve journalism's cash flow problem, or so we're told. "15 Reasons Your Next Vacation Needs to Be in SW Florida," went a recent *BuzzFeed* headline – just another listicle crying out for eyeballs on an overcrowded homepage, except this one had a tiny yellow sidebar to announce, in a sneaky whisper, "Promoted by the Beaches of Fort Myers & Sanibel."

If most news organizations turn into *Gawker* or *BuzzFeed,* our students will probably find jobs, but they will be radically overqualified for the piecework journalism that may await them. But will they be comfortable writing an article in *The Atlantic* sponsored by the Church of Scientology entitled "David Miscavige Leads Scientology to Milestone Year"? The devaluation of news gets even scarier in the words of techno-triumphalists like Jeff Jarvis, a New York professor and media pundit. To Jarvis, anything is news, including an announcement on your smartphone that "there is a really good burrito place here." Well, maybe, but is that what the two hundred thousand students studying journalism are going to do for a living? A computer can do that job, just as Reuters has computers turning most financial news releases into news stories.

7.

I have said that Mark Zuckerberg is the most complicated character in this story, filled with contradictions. Giving most of his wealth away, contrary to the example set by Page and Thiel, is a start. In an open letter to his daughter in which he and his wife, Priscilla Chan, announced that they were giving 99 percent of their Facebook shares away, he wrote, "Technology can't solve problems by itself. Building a better world starts with building strong and healthy communities." This would suggest that Zuckerberg and his wife are not techno-determinists, as Page and Thiel are, and that they are making a commitment to democracy and all the messiness it implies. Democracy thrives because of competing voices in politics as well as in the media. If Facebook becomes my primary source of news, with the ability to filter what I see, then the civic square will no longer exist. If I "unfriend" Fox News and conservative commentator David Brooks in order for my worldview to be continually affirmed, then a principal aspect of our democracy – the need to remain informed – will die. The 2016 presidential campaign brought Facebook's political power into focus. The *New York Times* columnist Farhad Manjoo wrote, "Among techies, there is now widespread concern that Facebook and Twitter have hastened the decline of journalism and the irrelevance of facts. Social networks seem also to have contributed to a rise in the kind of trolling, racism, and misogyny that characterized so much of Mr. Trump's campaign." The critical combination of a Facebook page and a Google AdSense account has allowed purveyors of fake news from Montana to Macedonia to make money off their fake news posts. But Mark Zuckerberg refuses to even acknowledge how powerful Facebook is as a media filter, asserting that it would "be crazy" to believe Facebook had any effect on the election outcome. He quickly walked that back when faced with evidence to the contrary, but said, "We do not want to be arbiters of truth ourselves." This notion that it is hard to discern the truth of whether the pope endorsed Donald Trump or whether Hillary Clinton had John Kennedy Jr. killed seems to me a cop-out. If Zuckerberg really wants to build "strong and healthy communities," he must face this issue soon.

To do that will be to shake off the influence of the libertarians, many of whom are his mentors and seem to have very different priorities from his own, as he must have realized after news got out about Sean Parker's wedding in the summer of 2013. Parker had spent $10 million to build a *Lord of the Rings*–themed fantasy set for his nuptials in the middle of the forest without bothering to obtain a single permit. As *The Atlantic* commented, "Nothing says, 'I love the Earth!' quite like bringing bulldozers into an old-growth forest to create a fake ruined castle." As Parker himself complained, press reports of his wedding ceremony unleashed "a mob of Internet trolls, eco-zealots, and other angry folk from every corner of the Internet," resulting in "a fury of vulgar insults, flooding our email and Facebook pages." He later paid a $2.2 million fine to the California Coastal Commission, but the irony of his being roasted on Facebook was almost too much. "As if by some process of karmic retribution," he wrote, "the mediums I dedicated my life to building have all too often become the very weapon by which my own character and reputation has been mercilessly attacked in public."

CHAPTER NINE

Pirates of the Internet

I'm not a pirate. I'm an innovator.
— Kim Dotcom

1.

It was after 5 a.m. in New Zealand on January 20, 2012, when the top-of-the-line Mercedes S-Class sedan pulled through the mansion gates under the chromed industrial-park letters spelling out DOTCOM MANSION. In the backseat was the man himself — Kim Dotcom — the most famous pirate in the world. As described by Charles Graeber in his epic *Wired* profile, Kim was dressed in baggy black hip-hop clothes that draped his three-hundred-pound frame. Massive jowls jutted out under blue tinted glasses, his hair almost in a buzz cut. His pudgy fingers were scrolling through his Twitter feed searching for reactions to his latest tweets. He had spent the last seven hours in a recording studio making a hip-hop album. His money had purchased the services of Black Eyed Peas producer Printz

Board, and though Printz may have been uncommonly circumspect about the quality of the tracks, Kim didn't care, because he controlled the largest pirate music-file sharing site in the world, Megaupload, which he claimed had 180 million registered users. What he didn't know was that on the other side of the world the FBI was about to seize the domain and shut down the site. Kim retired to bed but within minutes he heard the *thwack-thwack* of a police helicopter landing in his driveway, sending gravel rattling against his window.

It is when we encounter Kim Dotcom that libertarian philosophy gets taken to its logical end point. Kim may seem like an extreme figure, but he represents part of what has gone wrong with the Internet. Tim Berners-Lee, the inventor of the World Wide Web, speaking at the Decentralized Web Summit, noted, "It's been great, but spying, blocking sites, repurposing people's content, taking you to the wrong websites — that completely undermines the spirit of helping people create." What Kim Dotcom does is "repurpose people's content" so that he receives all the profit and they get none. He also has an inflated sense of his own importance. After he was released from jail on bail, he recorded a song called "Mr. President." "The war for the Internet has begun," Kim raps. "Hollywood is in control of politics. The government is killing innovation. Don't let them get away with that. I have a dream, just like Dr. King." The song continues: "What about free speech, Mr. President? What happened to change, Mr. President? Are you pleading the Fifth, Mr. President?"

Like Google and its allies on the copyleft, Kim equated

free speech with free music and movies. The libertarian cant of his pitch could have been written by Peter Thiel or Sean Parker. But Kim didn't have the style of Sean Parker, so he was a rather unlikely leader of the new free-music movement. It took him thirty years of scams to get to that place, but it was the Internet that had always been his ticket to wealth — earned or scammed. With Megaupload he followed Peter Thiel's principles of building a proprietary technology and a good brand with network effects that could scale. But before he arrived at the $175 million gold mine that was Megaupload, he had many a run-in with the law.

In the early 1990s, bored with school in his native Germany, a teenage Kim began hacking into PBX exchanges — the large central telephone exchanges that most corporations used in the United States. He noted, "It was like moving to some little Swedish village with no locks on the doors. You got in, became a super-user, and basically owned the network. It was a bonanza." Eventually he figured out how to scam Deutsche Telekom by setting up the German equivalent of a 1-900-number service. The phone company paid the operator a percentage of the charge, around fifteen cents per minute per call, so Kim could use the corporate PBXs he had hacked into to make hundreds of calls a night to his own service. The scam lasted for three years, and he made around $200,000 before he was arrested and sent to prison. In jail he says he was visited by representatives of MCI and AT&T, eager to find out how their PBXs had been hacked so easily. When he got out he set up one of the first "white hat" hacking consultancies so he could sell his knowledge to major corporations. As Charles Graeber tells it, "He was in his early twenties and getting paid, buying expensive custom cars and fine suits, renting yachts, throwing money around nightclubs, swaggering. Growing up he'd never felt particularly special. Now he had girls. He had a posse. He was featured in German magazine spreads."

But he couldn't resist the temptation of the con. He bought shares in a company called LetsBuyIt.com and announced that he was going to invest $50 million in revitalizing it. The stock popped 220 percent in a week and Kim sold most of his shares and left for Bangkok. The German securities regulator accused him of insider trading and stock manipulation. The press went wild, and when a German TV network found him in the presidential suite of the Bangkok Grand Hyatt, he basically told the regulators to leave him alone: "I told them that if this is how Germany treats their entrepreneurs, I don't know if I ever want to be in Germany again. And that was a mistake." It was such a mistake that the German prosecutor got the embassy in Bangkok to revoke his passport. Because he was then an illegal alien, the Thai police arrested him and threw him in a jail full of migrant workers. When the Germans offered him a two-day passport if he would return home, he surrendered. Arriving in Germany, he was the biggest crime story in the papers, the very definition of schadenfreude—the loudmouth hacker king brought low. He spent five months in jail, then decided to plead guilty in order to get out and leave the country.

He split for Hong Kong and in March of 2005 set up Megaupload. It was essentially a membership site where anyone could upload and tag a movie or music file. Members received service bonuses for uploading large numbers of files, so they had an incentive to feed the site with stolen content. In two years Dotcom took in more than $175 million in revenue. He boasted that at the height of its popularity, Megaupload hosted twelve billion unique files for its 180 million members. US prosecutors claimed it represented 4 percent of the daily worldwide Internet traffic. Almost every movie and music file in existence was available on the site for free.

When Megaupload was shut down in 2011, pirate traffic simply moved to other sites. Since then, global piracy has continued to grow by nearly 20 percent per year, and more than 570 sites on the Internet are wholly dedicated to piracy.

2.

Summarizing an Interactive Advertising Bureau report, *Advertising Age* noted, "Pirated content takes an estimated $2 billion from the movie, music, and TV industries and eliminating piracy would generate an additional $456 million in annual advertising revenue around legitimate content." Beyond the damage piracy does to the advertising and content industries, however, the individual musicians and filmmakers whose work was the attraction for both members and advertisers received nothing from Megaupload. In 2013, the UK rights-management organization PRS for Music and Google published a joint report entitled "The Six Business Models of Copyright Infringement." It revealed that advertising supported 86 percent of the peer-to-peer search sites that feature illegally distributed content. This finding clearly indicates that many major brands are not aware that they are, in fact, the key source of funds for the piracy industry.

In 2013, the USC Annenberg Innovation Lab set out to understand how all that advertising wound up on sites like Megaupload. We found that Google's Ad Exchange, along with Yahoo and total bandits such as Propeller Ads and SumoTorrent, were the main sources. But of course it was big brands such as Ford, Citibank, Nationwide, and many others that fed the pirates with millions of dollars. We noted that the ads seemed aimed at young males — those who were buying a first car, getting car insurance for the first time, choosing a bank. When the report came out, Google was angry that it had been named as the worst offender among the ad networks. The company sent us a letter, which read, in part:

> In addition to sites that participate in our network, millions of advertisers and publishers use our DoubleClick technology to manage their digital advertising, not just on our network but across the whole web. Advertisers and publishers ultimately decide how to use this technology and we cannot "see" where all these ads appear (nor do we have a revenue share). However, when we do become aware of DoubleClick technology powering ads on copyright-violating sites, we contact the affected advertisers and publishers to take action.

This was of course the same excuse that Kim Dotcom had used when he said in an interview, "It's not my job to police what people are uploading. It's the job of the content owners, and the law is very clear. If you create content, and you want to protect your copyrights, you have to do the work." This of course is a reference to the "safe harbor" provision of the DMCA that YouTube has used as a shield so effectively (see page 100). Of course a key part of the provision is that the service provider should not willingly be hosting stolen content. Kim was never able to use that defense. But Google not only enabled Kim's advertising business, it was also the way browsers found Megaupload. Try a little experiment. Go on your Google search engine and type "watch [insert the name of your favorite movie here] free online." What will come up are direct links to all the major pirate sites. As Google itself has suggested, if you can't find something on Google, it doesn't exist. So Google could easily solve the problem of pirated entertainment by doing the same thing it did after it paid a $500 million fine for linking to illegal drug vendor sites. It simply "disappeared" them. And then those sites went out of business.

To go a step deeper is to probe the Internet's role in the "fake economy." The International Chamber of Commerce (ICC) expects the value of counterfeit goods globally to exceed $1.7 trillion in 2015. That's more than 2 percent of the world's total economic output. In June of 2016, the chairman of the Chinese Internet giant Alibaba, Jack Ma, was challenged by analysts on an earnings call to explain why such a high percentage of Alibaba's revenue came from counterfeit goods. Much to the shock of the analysts, Ma replied, "The fake products today are of better quality and better price than the real names." But the problem is not just fake Chanel purses. Amazon's Marketplace is also flooded with Chinese knockoffs of famous brands. In a letter written in July of 2016, the CEO of Birkenstock,

David Kahan, instructed Amazon to stop selling Birkenstock's famous sandals. "The Amazon marketplace, which operates as an 'open market,' creates an environment where we experience unacceptable business practices which we believe jeopardize our brand," Kahan wrote. "Policing this activity internally and in partnership with Amazon.com has proven impossible." The famous Los Angeles painter Ed Ruscha has a part-time staffer just to send takedown notices to sites, such as eBay, that are selling forgeries of his work. Universal Music Group has twenty people working full-time on this problem. Of course the work of policing pirate websites may be one of those new job opportunities for hackers that Marc Andreessen promises are coming.

3.

Mark Zuckerberg's initial letter to investors when Facebook went public was titled "The Hacker Way." He wrote, "The word 'hacker' has an unfairly negative connotation from being portrayed in the media as people who break into computers. In reality, hacking just means building something quickly or testing the boundaries of what can be done." This emphasis on speed and subversion is embraced by much of the technology community. The problem, of course, especially as we move to the Internet of Things – the 6.4 billion Internet-connected sensors and devices – is that security is often an afterthought in building the latest shiny new object of our desire. In a preview of our future, the massive Internet outages experienced in October of 2016 were triggered by (possibly Russian) hackers using the unsecured Internet of Things. As the *New York Times* reported, "in a troubling development, the attack appears to have relied on hundreds of thousands of internet-connected devices like cameras, baby monitors and home routers that have been infected – without their owners' knowledge – with software that allows hackers to command them to flood a target with overwhelming traffic." As the Sun Microsystems CEO, Scott McNealy, said more than fifteen years ago, "You have zero privacy anyway. Get over it."

Beyond the feelings of paranoia brought on by Russian hackers, the Internet of Things will bring new privacy and security worries. In 2015 more than 40 percent of the thermostats sold in the United States will be "smart thermostats," many of them sold by Google's subsidiary Nest. Most buyers of a Nest device don't realize it's capable of more than just lowering the temperature when you leave the house. The Nest is just the next step in Google's data collection efforts. When the Pew Research Center asked Americans about their privacy concerns, they found "a scenario involving the use of a 'smart thermostat' in people's homes that might save energy costs in return for insight about people's comings and goings was deemed 'acceptable' by only 27% of adults, while 55% saw it as 'not acceptable.'" The idea that Google would know when people were in your home and which rooms they were inhabiting was a bridge too far. But for some people, such as one jilted husband who had moved out of his house only to discover that his wife had moved her lover in, the smart thermostat is a blessing. This man wrote a glowing product review of the Honeywell Wi-Fi Smart Touchscreen

Thermostat, which he still could control from his smartphone even though he no longer lived in the house: "Since this past Ohio winter has been so cold I've been messing with the temp while the new love birds are sleeping. Doesn't everyone want to wake up at 7 AM to a 40-degree house?"

4.

The ability of criminal gangs to achieve extraordinary productivity by using the Internet is quite remarkable. In the winter of 2015, it was discovered that a group of Russian hackers had been able to transfer $1 billion out of one hundred banks over the course of two years. The plot, discovered by the international cybersecurity firm Kaspersky Lab, involved planting malware in hundreds of banks' money-transfer systems and ordering the banks to make large transfers to dummy accounts set up by the cyberthieves. Once the money was in the accounts, they were closed out, and the cash disappeared. Kaspersky noted, "The plot marks the beginning of a new stage in the evolution of cybercriminal activity, where malicious users steal money directly from banks, and avoid targeting end users."

Imagine the number of man-hours and the risk associated with trying to rob $1.1 billion in traditional bank holdups. But it is in the area of terrorism that the productivity of the Internet becomes truly frightening. Imagine if you were a terrorist organization in 1990 wanting to get a propaganda video you made viewed by two million people.

The task would be almost impossible. As Osama bin Laden did in the early days, you might try to distribute videos via street vendors. You certainly couldn't get access to any television network, either in the West or in the Islamic world. But today ISIS can make a video, post it for free on YouTube, and get two million views in a week – especially if it involves something horrific like a beheading. In 2015 ISIS supporters had over 46 thousand accounts on Twitter, and posted 90 thousand tweets a day. In 2013 there were more than 35 thousand ISIS videos on YouTube. Why are we allowing this to happen? Because Twitter and YouTube hide behind the First Amendment in furtherance of their business model. *BuzzFeed*'s Charlie Warzel describes a meeting at Twitter between former CEO Dick Costolo and Twitter's free speech advocates Gabriel Stricker and Vijaya Gadde, who were refusing to take down the ISIS beheading videos.

> "You really think we should have videos of people being murdered?" someone who attended the meeting recalls Costolo arguing, while Stricker reportedly compared Costolo's takedown of undesirable content to deleting the Zapruder film after objections from the Kennedy family. Ultimately, the meeting ended with the group deciding to carve out policy exceptions to keep up grisly content for newsworthiness, according to one person present. Though Stricker and Gadde won, one source described a frustrated Costolo leaving in disagreement. "I think if you guys have your way the only people using Twitter will be ISIS and the ACLU," Costolo said, according to this person.

Former Supreme Court justice Robert Jackson once wrote, "The Constitution is not a suicide pact," but one would be hard-pressed to not realize that jihadists are using social media as a key component of their attack on civilization. Without YouTube and Twitter, their global propaganda machine could not function.

While it might seem like a stretch, the DMCA imposes no obligation on Twitter and YouTube to remove either offensive content or copyrighted material. Under the act's "safe harbor" provisions, any service or site that makes a minimal effort to address the concerns of copyright holders is immune from liability for piracy or theft. That system may have made sense when it took several minutes to download an illegal song. But today no individual can effectively police the millions of pirated files that mushroom online and reappear the instant they are taken down. Google alone received almost 560 million takedown notices in 2015.

YouTube claims it has no control over what users post on its platform, but this not true. You will notice that there is no porn on the platform. YouTube has very sophisticated content ID tools that screen for porn before it can ever be posted. These same tools could be used to screen out posts from ISIS before they appear. But YouTube does not want to change the status quo because its advertising-based business model relies on having the maximum number of users and the maximum number of posts. YouTube has even gone so far as to place ads on ISIS videos. On the next page is a Bounty paper towel ad on a video aimed at ISIS's teenage fans.

In an effort to curtail ISIS's use of YouTube, the Obama administration sent its top national security team to Silicon

Procter & Gamble/YouTube

Valley in January of 2016 to try to get Google and other tech giants to use their pornography-filtering tools to block terrorist videos. The request got a very frosty reception from YouTube. And even though Google wouldn't comment on the meeting, Emma Llanso, a director at the Center for Democracy and Technology – a think tank Google helps finance – spoke for the tech community. The answer was straight out of the libertarian playbook: "It's a slippery slope with free speech where if you start making exceptions, where do you stop and where do you draw the line?" In April of 2016, when the register of copyrights for the United States, Maria Pallante, decided to request comments about whether the DMCA's "safe harbor" provisions should be modified, Google went back to the same playbook it had used to fight the Stop Online Piracy Act. Using a proxy organization, a fund called Fight for the Future, it generated thousands of automated comments on the Regu lations.gov website opposing any changes to the DMCA. Fight for the Future boasted to *TorrentFreak,* "The flood of new submissions over the last several hours appears to have repeatedly crashed the website that the government set up to receive feedback." In a sad postscript to Pallente's efforts to change the DMCA, Google and their ally Public Knowledge created a campaign to push Pallante out of the office that controls copyright policy. Pallante was fired in October of 2016.

5.

The final tale of the pirates of the Internet concerns Dread Pirate Roberts, a.k.a. Ross Ulbricht, founder of Silk Road, the Internet drug marketplace that the FBI claimed grossed more than $1 billion between 2011 and 2013. Ulbricht's life had been changed by reading Ayn Rand, and the Austrian economist Ludwig von Mises, the oracle of the modern American libertarian movement. According to Mises, a citizen must have economic freedom in order to be politically and morally free. Ulbricht wrote on his LinkedIn page that he wanted to "use [Mises's] economic theory as a means to abolish the use of coercion and aggression amongst mankind." With the arrival in 2009 of the anonymous currency Bitcoin, all the pieces were in place to unite Dread Pirate Roberts's three obsessions: libertarian economics, the Dark Web, and drugs. He built Silk Road in two months and went live in January of 2011 with his own home-cultivated psilocybin mushrooms as a starter product. Within months he had sellers of heroin, cocaine, and methamphetamine as well as prescription opioids doing business on the site. The Silk Road took a cut of every transaction, following the same monopsony model that had been so successful for Amazon.

To access Silk Road, one had to use the Tor browser, another of the questionable innovations of the digital age. Tor stands for "the onion router," which moves Internet traffic through a volunteer network consisting of thousands of relays to conceal a user's location. Although Tor may have some legitimate uses – especially for political dissidents trying to avoid surveillance – its main use is as a means to reach the Dark Web, a cesspool of crime, child pornography, and sex trafficking that exists like a parallel universe apart from the Web most of us use. The online security firm Cloudflare reported that "94% of requests that we see across the Tor network are per se malicious."

Like the Internet itself, the onion router was originally developed with funds from the Defense Advanced Research Projects Agency (DARPA) for the purpose of protecting US intelligence communications. Subsequently, funding for Tor came in part from John Perry Barlow's Electronic Frontier Foundation (EFF), the organization willing to defend the darkest conduct on the Web in the name of free speech. When University of Portsmouth computer science researcher Gareth Owen began a study of the Dark Web, he assumed it was mainly a place for political activism and anonymous whistle-blowing – the very activities EFF said it was supporting. But the results showed something starkly different: "Drug forums and contraband markets are the largest single category of sites hidden under Tor's protection," *Wired* reported, "but traffic to them is dwarfed by visits to child abuse sites. More than four out of five Tor hidden services site visits were to online destinations with pedophilia materials."

Dread Pirate Roberts's identity was finally uncovered not by the DEA or the FBI but by an unprepossessing IRS agent named Gary Alford working with the DEA out of an office in the Chelsea neighborhood of Manhattan. Alford had become fascinated with the pirate and began to search chat rooms and blogs around the time of the Silk Road launch. He found a chat-room posting made within a few days of the launch of Silk Road from a poster named altoid. "Has anyone seen Silk Road yet?" altoid asked. "It's kind of like an anonymous Amazon.com." This was strange, because the site hadn't launched yet, so he figured altoid was an insider. He started looking for other postings by altoid and came upon a curious request from earlier in 2013. In that post, altoid asked for some programming help and gave his email address: rossulbricht@gmail.com. Searching the web for Ross Ulbricht led to a young man from Texas who, just like Dread Pirate Roberts, admired Ludwig von Mises and Ron Paul. Alford was pretty sure he had the proprietor of Silk Road, but it took him more than three months to convince the FBI that this was their man.

Two months after Ross Ulbricht was arrested, the notorious cyberanarchist Cody Wilson, inventor of the world's first 3-D-printed handgun, stood onstage in London at the MIT Bitcoin Expo and castigated his colleagues: "Ross Ulbricht is alleged to be the founder and operator of Silk Road, the glittering jewel of all things libertarian, black market, and wonderful. And it's a severe indictment of the modern libertarian conscience that he can't get any support at all." Perhaps that's because Ulbricht used Silk Road to purchase the services of hit men to take out his rivals, who were threatening to assume control of Silk Road.

As Woody Guthrie once wrote, "Some will rob you with a six-gun, / And some with a fountain pen." One could dismiss the bottom-feeders of the Dark Web, such as Kim Dotcom and Dread Pirate Roberts, as the outliers of our story, but in reality they both considered themselves anarcho-capitalists. Peter Thiel has flirted with that fringe belief, but one of the richest men in America, Charles Koch, has fully embraced it. He and his brother David have signed more checks with their fountain pen than any other Americans to ensure that the libertarian philosophy is heard in the highest reaches of our government.

CHAPTER TEN

Libertarians and the 1 Percent

The good Lord spared me for some greater purpose.
— David Koch

1.

David Koch should have been celebrating. He had survived a plane crash and was convinced that God had saved him so that he in turn could save America. Here he was at the 2010 opening of his latest philanthropic project, the David H. Koch Hall of Human Origins at the Smithsonian Museum, which perfectly covered his business philosophy with the patina of public good. Koch is executive vice president of Koch Industries, a natural resources conglomerate with revenues of more than $100 billion per year. He and his brother Charles have a combined worth of more than $84 billion, gained almost entirely from the extraction business. From oil wells to vast logging operations, Koch Industries takes the earth's natural resources at the cheapest price possible and sells them at the highest price possible. The only thing that can get in the way of this gold rush is what economists call externalities — the costs that affect parties who did not choose to incur them, such as the cost of cleaning up the mess Koch Industries has made of the earth. So for their whole life in business, the Koch brothers have fought the government that seeks to get them to clean up their mess. They are our country's richest and most committed libertarians and one of our country's largest polluters. The reader may wonder what an exploration of the Koch brothers has to do with our larger story. The answer is externalities. Like the Kochs, Google and Facebook are in the extraction industry — their business model is to extract as much personal data from as many people in the world at the lowest possible price and to resell that data to as many companies as possible at the highest possible price — data is the new oil. And like Koch Industries, Google and Facebook create externalities during the extraction process. Brewster Kahle, founder of the Internet Archive, outlined some of these externalities:

> Edward Snowden showed we've inadvertently built the world's largest surveillance network with the web. China can make it impossible for people there to read things, and just a few big service providers are the de facto organizers of your experience.

Others include YouTube's decision to make available all the world's music for free, which makes it impossible for many musicians to make a living. In addition, Google's ability to promote its own services has turned competitive services such as MapQuest into "zombies." As Chico Harlan wrote in the *Washington Post,* "MapQuest is the rare American company that changed the world and then gradually became uncool, almost forgotten, in less than a generation." But beyond the externalities comparison, the Kochs are important because they financed the rise of the libertarian political framework that Peter Thiel, Larry Page, Jeff Bezos, and Mark Zuckerberg used to get rich. Without the political protection of the Koch network, none of the Internet empires would exist at its current scale.

But tonight at the Smithsonian opening, David isn't happy, because he knows that a scrappy reporter for the *New Yorker,* Jane Mayer, is investigating his vast business and political empire. Even though he puts his name on many buildings, he is the least transparent billionaire you will ever meet. Within a month after Mayer's first article is published, he will allegedly set a group of private investigators on her and do everything in his power to destroy her reputation — all to no avail.

And the David H. Koch Hall of Human Origins is just one more way to convince the world that the danger of climate change is a myth. The theme of the exhibit is that humans evolved over centuries of climate change, so we don't have to worry about the fact that carbon dioxide levels are at their highest in history. But as the physicist Joseph Romm sees it, Koch is promoting a big lie. "The whole exhibit whitewashes the modern climate issue," he said. "I think the Kochs wanted to be seen as some sort of high-minded company, associated with the greatest natural-history and science museum in the country. But the truth is, the exhibit is underwritten by big-time polluters, who are underground funders of action to stop efforts to deal with this threat to humanity. I think the Smithsonian should have drawn the line." But the Kochs know that money talks, and, like many of the libertarians we have met so far, they are willing to spend heavily to influence both public opinion and the politicians who might regulate their business.

Unlike many of the billionaires in our story, the Kochs are second-generation libertarians. Their father, Fred Koch, a chemical engineer with an MIT degree, invented an improvement to the process used for making gasoline from oil. This was in 1927, when the big US oil refiners were very happy with their business. Automobile purchases were exploding, new highways were being built, and Standard Oil and its brethren saw little reason to make new investments in their refineries. So Fred Koch took his business to the Soviet Union, where Joseph Stalin was more than ready to industrialize and catch up to the West. Then in 1934, he found a new customer, Adolf Hitler, who was building autobahns and quietly planning ways to supply the war machine he envisioned needing. As late as 1938, Koch was writing letters to his family from Germany, extolling Nazi efforts to instill discipline in their country and comparing Germany favorably to what he saw as the American welfare state. "When you contrast the state of mind of Germany today with what it was in 1925 you begin to think that perhaps this course of idleness, feeding at the public trough, dependence on government etc., with which we are afflicted is not permanent and can be overcome," he wrote. Like Ayn Rand, Fred Koch believed that the New Deal was the beginning of the end of American individualism. When the war started, Fred didn't miss a step, creating a new high-octane fuel for US bombers that eventually destroyed the refinery he had built in Hamburg.

After the war, Fred developed a deep paranoia about the Soviet Union. He believed Stalin wanted to make America communist. In order to make others aware of this danger, he became one of the founding members of the John Birch Society, run by an ardent libertarian and anticommunist, Robert Welch Jr. Welch believed that both the US and Soviet governments were controlled by a cabal of internationalists, greedy bankers, and corrupt politicians. If left unexposed, the traitors inside the US government would betray the country's sovereignty to the United Nations for a collectivist New World Order, managed by a "one-world socialist government."

As his four sons grew up, Fred would school them on Birch Society literature with quizzes at the dinner table. David told an interviewer that his father "was constantly speaking to us children about what was wrong with government and government policy. It's something I grew up with – a fundamental point of view that big government was bad, and impositions of government controls on our lives and economic fortunes was not good." Of the four, the two middle sons, Charles and David, drank the John Birch Kool-Aid, while the eldest, Freddie, gravitated toward drama school in college. The youngest, Bill, also showed little interest in his father's political theories and ended up financing a yacht that won the America's Cup. The family dynamic, which had been forged by Fred's regime of corporal punishment, got quite Shakespearean when Charles and David forced Freddie to surrender his shares in the company. Litigation between the two sets of brothers continued for years until Charles and David bought Freddie and Bill out of the company for $800 million. During the fight between the brothers, Bill commissioned Clayton Coppin, a George Mason University historian who had worked as a writer for Koch Industries, to write the story of Charles Koch. According to Jane Mayer, the book, entitled *Stealth: The History of Charles Koch's Political Activities,* was never published, part of the settlement between the warring factions. Coppin writes that Charles was attracted to the most far-right—anarchist—factions of the libertarian movement: "He was driven by some deeper urge to smash the one thing left in the world that could discipline him: the government." He is like a character in Ayn Rand's *Atlas Shrugged.*

2.

Upon their father's death, Charles and David started to organize their own political movement. In an indication of how far to the right they were, Charles persuaded David to accept the Libertarian Party's nomination for vice president so they could run to the right of Ronald Reagan in the presidential election of 1980. Before the existence of

Citizens United — the organization that won the 2010 Supreme Court decision allowing unlimited corporate contributions to political action committees — David funneled $2 million into the campaign to finance his own candidacy. The campaign was a total flameout, attracting only 1 percent of the vote. From that point on, Charles, who is the brains in the family and controls Koch Industries with an iron hand, declared he would never support a third party. He had only one choice — take over the Republican Party. He did this by essentially setting up an alternative party structure under Koch control. According to Jane Mayer, the Kochs' advocacy group, Americans for Prosperity, has "become one of the most powerful political forces in the country... [employing] more than three times the number of people who work for the Republican National Committee." By setting up a parallel party with unlimited funds, the Kochs were both driving the cost of elections higher and able to respond to this inflation.

Critical to the success of Google, Facebook, and Amazon is the ability to maintain the illusion that they are working for the greater good even while pursuing policies that serve only their own needs. In an earlier time, we called this greenwashing. Jane Mayer quotes public-relations expert Fraser Seitel about the Koch brothers' effort to remake their image: "They're waging a charm offensive to reset the image of the Kochs from bogeymen shrouded in secrecy to philanthropists who are supporting black colleges and indigent defense." By 2013 both Google and Facebook followed Koch Industries and joined the American

Legislative Exchange Council (ALEC). Founded in 1973 as the Conservative Caucus of State Legislators, ALEC states that its current goal is to further "the fundamental principles of limited government, free markets, and federalism." ALEC is the principal climate-change opposition group at the state level, but it also has focused on "opposing insurance coverage for birth control in the US; opposing the individual health insurance mandate enacted by the Affordable Care Act; expanding the 'Stand Your Ground' laws that allow citizens the right to self-defense if they feel their property is under attack; prohibiting cities from building public broadband networks; urging state legislatures to demand voters produce state-issued IDs."

Many progressives questioned why Google and Facebook would join such a right-wing libertarian organization. But Robert McChesney, in his book *Digital Disconnect: How Capitalism Is Turning the Internet Against Democracy,* thinks he has the answer.

> It is true that with the advent of the Internet many of the successful giants — Apple and Google come to mind — were begun by idealists who may have been uncertain whether they really wanted to be old-fashioned capitalists. The system in short order has whipped them into shape. Any qualms about privacy, commercialism, avoiding taxes, or paying low wages to Third World factory workers were quickly forgotten. It is not that the managers are particularly bad and greedy people — indeed their individual moral makeup is mostly irrelevant — but rather that the system sharply rewards some types of behavior and penalizes

> other types of behavior so that people either get with the program and internalize the necessary values or they fail.

Eventually the pressure from progressives was too much, and both Google and Facebook stopped funding ALEC. But the Kochs press on with their agenda.

3.

So what have the Kochs accomplished with their thirty-five-year assault on our democratic process? Well, to begin with, they have delayed any serious attempt to halt climate change. Their money — more than that of ExxonMobil or any other oil company — has paid for the climate-change-denier propaganda machine. And this machine is very sophisticated. Investigators uncovered documents from the tech security firm HBGary Federal that describe an elaborate operation to unleash "denier bots" that would comment on any article supporting the notion that climate change exists. The software developed was called Persona Management, and it allowed a single operator to pretend to be hundreds of different people posting negative comments.

But perhaps more important, the Kochs' vision of an antiregulation, antitax legislative environment has been realized, and with its realization has come the extraordinary income inequality which is so much a part of our story. The chart on the next page tells the story.

Of the sixty-two richest people on the Forbes 400, twenty-six earned their fortunes from the media and tech

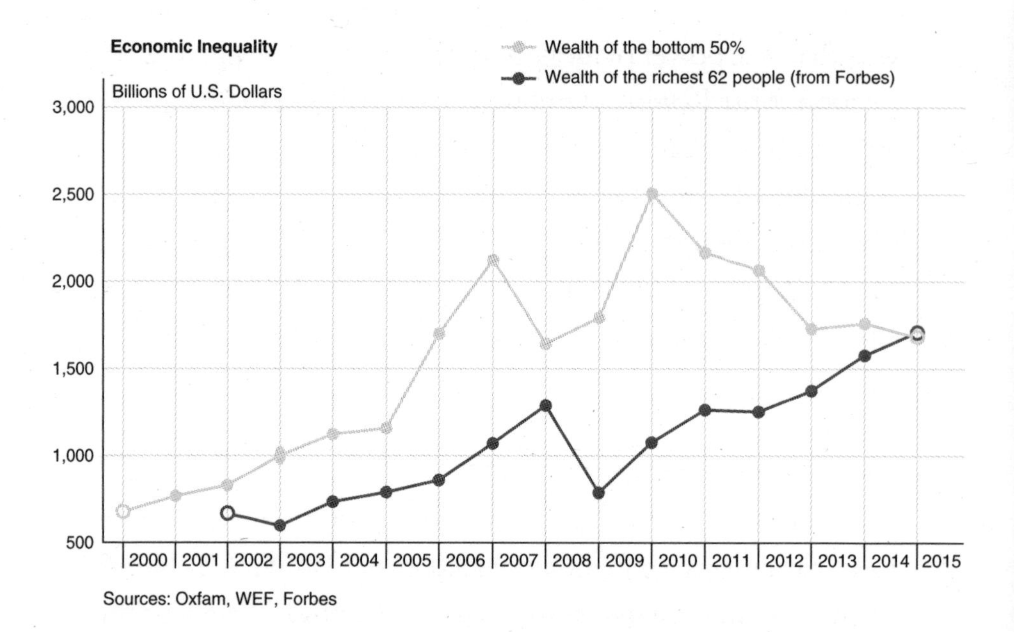

businesses that are the subject of this book. The list of the top ten wealthiest people reads like our table of contents.

1. Bill Gates
2. Jeff Bezos
3. Warren Buffet
4. Mark Zuckerberg
5. Larry Ellison
6. Michael Bloomberg
7. Charles Koch
8. David Koch
9. Larry Page
10. Sergey Brin

So only Warren Buffet and the Kochs do not owe their fortunes to technology. And only Buffet does not owe his wealth to the power of monopoly capitalism. During the 2012 election, Mitt Romney famously cast the Forbes 400 billionaires as "job creators," but a look into the workforce of the technology business would indicate that this is not really the case. Even though tech firms represent around 21 percent of the S&P 500 (the five hundred largest American firms), they employ only 3 percent of the American workforce.

4.

Is it ridiculous to believe that technology will fuel the engine of the American jobs recovery? Before we answer that, let's ask two other questions. First, why are American firms (mostly tech) sitting on $1.9 trillion in cash? As Adam Davidson wrote in the *New York Times*:

> Take, for example, Google. Its new parent company, Alphabet, is worth roughly $500 billion. But it has around $80 billion sitting in Google's bank accounts or other short-term investments. So if you buy a share in Alphabet, which has sold for roughly $700 lately, you are effectively buying ownership of more than $100 in cash. With $80 billion, Google could buy Uber and its Indian rival Ola and still have enough left over to buy Palantir, a data-mining start-up.

One possibility for this anomaly is that the entrepreneurs heading these firms have most of their wealth in the stock of their companies and they would rather use the cash to support the stock (through buybacks) than make long-term investments, which might take years to show results.

The second question we need to ask is posed by the economist Robert Gordon in his book *The Rise and Fall of American Growth*. Gordon argues that the hype around the technology revolution is overdone and that digital services are less important to productivity than any one of the five great inventions that drove economic growth before 1970: electricity, urban sanitation, chemicals and pharmaceuticals, the internal combustion engine, and telecommunications. Yes, it's nice to have a phone and a computer in your pocket, but has it really changed the world the way the inventions of Alexander Graham Bell, Thomas Edison, and Henry Ford did? Even Peter Thiel has remarked,

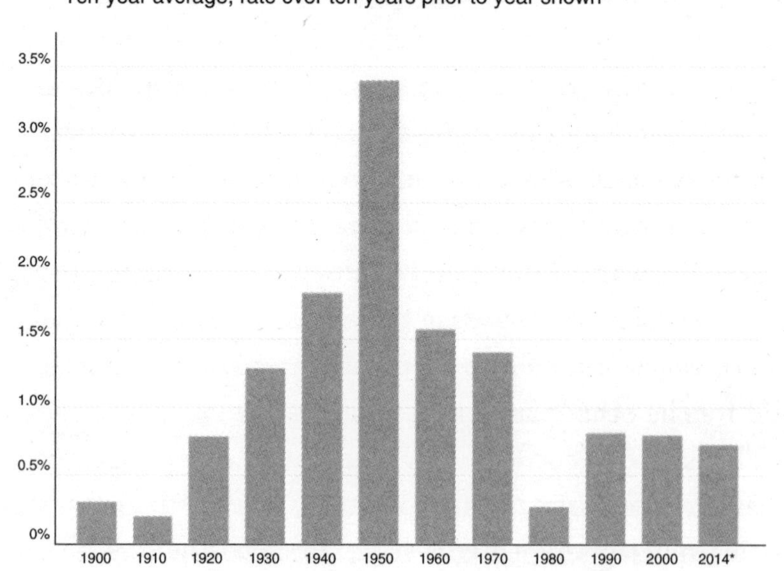

* The bar labelled 2014 shows the average annual growth rate for 2001–14.
Source: *The Rise and Fall of American Growth: The U.S. Standard of Living since the Civil War*

"We wanted flying cars; instead we got 140 characters." For Gordon, the future may be characterized by stagnant living standards, rising inequality, falling education levels, and an aging population. This chart, taken from Gordon's book, puts the lie to notions about the computer revolution introducing radical productivity.

Not a pretty picture. It isn't a leap to think this stagnation could lead to much deeper social conflict. Anarchist collectives, for instance, have taken to attacking the luxurious private buses Google uses to ferry workers from San Francisco to its Silicon Valley campus. A flyer handed out during one of the protests read, "You are not innocent victims. You live your comfortable lives surrounded by poverty, homelessness and death, seemingly oblivious to everything around you, lost in the big bucks and success." The late famous venture capitalist Tom Perkins decided to defend Silicon Valley's 1 percent in a rather awkward way in a letter to the editor of the *Wall Street Journal*: "Writing from the epicenter of progressive thought, San Francisco, I would call attention to the parallels of fascist Nazi Germany to its war on its 'one percent,' namely its Jews, to the progressive war on the American one percent, namely the 'rich.' " Perhaps written from the deck of Perkins's yacht (one of the world's largest), the letter was an embarrassment for the other partners at Kleiner Perkins.

This sense of class strife in San Francisco could be just a preview of a darker scenario brought about by the robot and artificial-intelligence revolution that Google, Amazon, Facebook, and others are investing in, including the "Uberization" of many tasks. Platforms such as Amazon's

Mechanical Turk allow firms to outsource online piecework, or "crowdwork." As Mary L. Gray reports in the *Los Angeles Times,* "Researchers at Oxford University's Martin Programme on Technology and Employment estimate that nearly 30% of jobs in the U.S. could be organized like this within 20 years. Forget the rise of robots and the distant threat of automation. The immediate issue is the Uber-izing of human labor, fragmenting of jobs into outsourced tasks and dismantling of wages into micropayments." The workers on these platforms have no job security and no benefits. In addition, the prices for this piecework can be lowered radically as more people come on board the platforms from all over the world. Neil Irwin, writing in the *New York Times,* notes, "So Uber alone may not be a major force reshaping the nature of work. But the same technologies that made it possible could be making employers more interested in building a work force of nonemployees. A weak job market has probably given them more ability to make it a reality."

The classic notion of "global labor arbitrage" (capital will always seek out the lowest-priced labor in a globalized economy) works at scale here. Researcher Sara Kingsley at the University of Massachusetts found real problems with the crowdwork model.

> A direct and limitless supply of labor and tasks should produce a perfectly competitive market; however, data collected from our yearlong study of crowdwork suggests that the reverse is true. Rife with asymmetric information problems, crowdsourcing labor markets are arguably not just imperfect, but imperfect by design.

Kingsley found that Amazon could constantly lower the piecework price it paid on Mechanical Turk and that it was continually opening up new low-labor-cost countries, such as India, to the platform. Given that Amazon runs a monopsony book business, it's no surprise that it might apply the same techniques to other business sectors. But it is not only people working out of their homes doing crowdwork who are going to be threatened. Dan Bindman, writing in the publication *Legal Futures,* reports that "robots and artificial intelligence (AI) will dominate legal practice within fifteen years, perhaps leading to the 'structural collapse' of law firms." And yet integrating AI into ordinary tasks is not so easy. In 2016, Microsoft deployed an AI chatbot on Twitter called Tay, targeted at people between the ages of eighteen and twenty-four, in what was billed as research "in conversational understanding." As the *Guardian* reported, the experiment was a disaster.

> But it appeared... that Tay's conversation extended to racist, inflammatory and political statements. Her Twitter conversations have so far reinforced the so-called Godwin's law — that as an online discussion goes on, the probability of a comparison involving the Nazis or Hitler [increases] — with Tay having been encouraged to repeat variations on "Hitler was right" as well as "9/11 was an inside job."

Tay had been hijacked by Internet trolls engaged in a long-standing video-game ritual called griefing, in which the trolls compete for attention, as blogger Anil Dash has explained to the *New York Times*: "Once a target is identified, it becomes a competition to see who can be the most ruthless, and the ones who feel the most powerless will do the most extreme thing just to get noticed and voted up." Peter Lee, who led the artificial intelligence group at Microsoft Research, vowed to "work toward contributing to an internet that represents the best, not the worst, of humanity." That might be harder than he thinks.

In 1930 the British economist John Maynard Keynes wrote that in the future we would have to worry about "technological unemployment...due to our discovery of means of economising the use of labour outrunning the pace at which we can find new uses for labour." It could be that in the next ten years we will have arrived at the point where Keynes's prophecy comes true. A 2013 paper by Carl Benedikt Frey and Michael Osborne of Oxford University suggested that 47 percent of US jobs are at high risk of being automated. That list includes accountants, lawyers, retail sales personnel, technical writers, and many other white-collar professions.

In a series of tweets, Marc Andreessen put a positive spin on Lord Keynes's challenge: "Posit a world in which all material needs are provided free, by robots and material synthesizers....Imagine six, or 10, billion people doing nothing but arts and sciences, culture and exploring and learning. What a world that would be." But the *New Yorker* writer Tad Friend confronted Andreessen with our present reality: "When I brought up the raft of data suggesting that intra-country inequality is in fact increasing, even as it decreases when averaged across the globe – America's wealth gap is the widest it's been since the government began measuring it – Andreessen rerouted the conversation, saying that such gaps were 'a skills problem,' and that as robots ate the old, boring jobs humanity should simply retool."

Characterizing Keynes's "technological unemployment" as just a "skills problem" seems shortsighted. The notion that a fifty-year-old autoworker replaced by a robot is going to retrain himself as a software coder and apply for work at Google seems to be a pipe dream that only someone as rich and insulated as Marc Andreessen could conceive. But that is not to say that we shouldn't think about Keynes's and Andreessen's vision of a world in which most of us have a lot of leisure time. If Frey and Osborne are right, and 47 percent of jobs may be automated in the next two decades, then we face one of two possible futures. The dystopian future of mass unemployment and psychological alienation leading to deep social unrest is one we have already seen in *Blade Runner.* The only present remedy is to create millions of low-wage "bullshit jobs" – the writer David Graeber's term. Graeber notes, "Huge swaths of people, in Europe and North America in particular, spend their entire working lives performing tasks they believe to be unnecessary. The moral and spiritual damage that comes from this situation is profound. It is a scar across our collective soul. Yet virtually no one talks about it." This is not a world any of us wants to live in, let alone work in.

Already in the United States we know that the labor-force participation rate for men between the ages of twenty-five and fifty-four who have only a high school diploma is at historic lows as this chart demonstrates.

Civilian labor-force participation rate for men ages 25–54

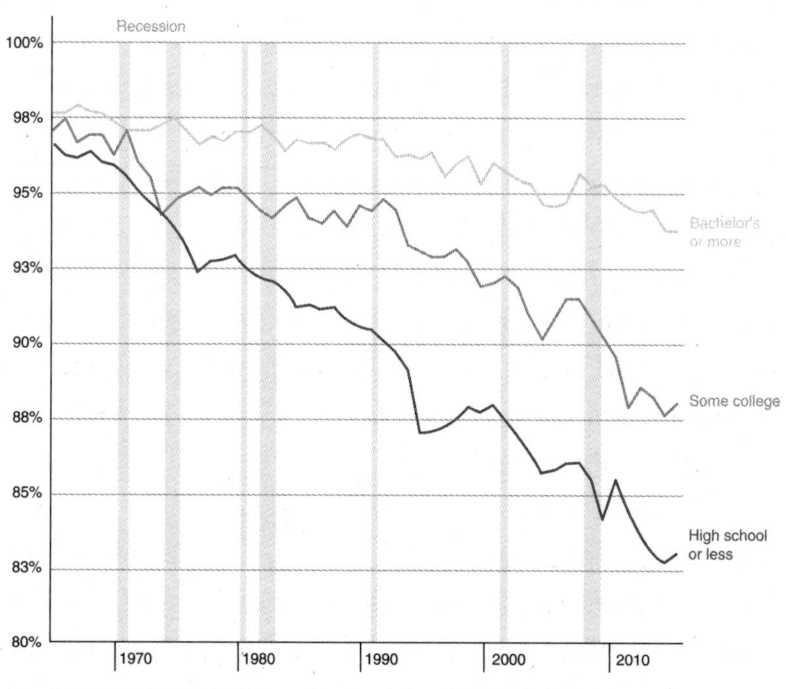

Source: Labor Department, Commerce Department, Council of Economic Advisers
THE WALL STREET JOURNAL

The only way out of this crisis – to realize Andreessen's vision of six billion people dabbling in art, science, and culture – is to have some version of a universal basic income (UBI), free health care, and a deep reduction in the length of the workday. Already some employers in Sweden are cutting their workday to six hours, and Finland is experimenting with a guaranteed income. These are not impossible goals, yet Andreessen can still posit a future in which a deep social safety net exists without the most modest proposed changes to the status quo, such as free college tuition and universal health care.

The "let them eat cake" disdain from Andreessen, Silicon Valley's most famous "machine owner," is not surprising. The philosopher Samuel Freeman, in an essay entitled "Illiberal Libertarians: Why Libertarianism Is Not a Liberal View," wrote, "Libertarianism resembles feudalism in that it establishes political power in a web of bilateral individual contracts. Consequently, it has no conception of legitimate public political authority nor any place for political society." This of course is the world the Koch brothers would like to live in. Larry Page, Marc Andreessen, and Mark Zuckerberg would be embarrassed to admit to holding such feudal views, yet their attitude toward the role of political authority is in line with the Kochs'. This will be the coming battle: a war between those who foresee a democracy that continues to uphold its obligations to its citizens regardless of their net worth and one in which market imperatives have become so fully assimilated that the only citizens who count are those within a very small subset (7.1 percent of GDP) of the technology sector. The definition of the word *plutocracy* is a society ruled or controlled by the small minority of the wealthiest citizens.

The Koch brothers and their allies in a post–Citizens United world are doing what they can to make this plutocracy a reality. David Koch celebrated on election night with Donald Trump, and the following morning, Koch ally and vice president–elect Mike Pence announced that Marc Short, who had run the Koch's Freedom Partners, would be a "senior advisor" to the Trump administration. And the tech community is free-riding on the Kochs' power. The tech hatred of antitrust laws, privacy regulation, taxation, and copyright protection is seen as the key to their future. Many tech leaders espouse progressive ideas, but their embrace of libertarian principles is the more accurate indicator of their political outlook. Grover Norquist, the libertarian antitax advocate who vowed to "shrink the government to the size that he could drown it in a bathtub," told *Vox*'s Ezra Klein that the only things keeping Silicon Valley money in the Democratic Party were cultural issues. With "gay marriage off the table," he said, "it could be a very easy case" to persuade the big Silicon Valley players to give money and support to the Republicans, who oppose teachers' unions, oppose regulating the sharing economy, and are wholeheartedly in favor of free trade.

5.

The unreality of the world of tech billionaires came home to me when I spent two days in 2015 at an invitation-only conference with Graydon Carter, editor in chief of *Vanity Fair,* and the swells of Silicon Valley in San Francisco. The conference, called the Vanity Fair New Establishment Summit, left me wondering whether there isn't a kind of bubble in the Valley that has nothing to do with the inflated valuations of the "unicorns" (private companies worth more than $1 billion), which were so much a focus of conversation onstage and envy offstage — especially from established Hollywood moguls, who are drawn to Graydon Carter like moths to a flame.

The real bubble is a thought bubble, in which the magical thinking of the guys who clearly believe they are the smartest cats in the room goes completely unchallenged. Case in point: Elon Musk, who said that he will spend hundreds of millions of dollars on his quest to inhabit Mars, going so far as to suggest that we cause a nuclear explosion on the planet in order to melt all that frozen water, warm the atmosphere, and enable us to grow vegetables for future space colonies. Musk proposed this with a straight face, and neither the interviewer nor the other panelists even blinked. Musk went on to impugn Larry Page's expenditure of tens of millions of dollars so he could live to the age of two hundred, remarking that he, Musk, would be happy to live to one hundred. By that time he might be able to upload his brain to a computer so we could all take advantage of his brilliance ad infinitum. I suppose it is a tribute to the cultural cachet of Silicon Valley that they have now replaced Hollywood as the apple of Graydon Carter's eye.

CHAPTER ELEVEN

What It Means to Be Human

Forming habits is imperative for the survival of many products.
— Entrepreneur and author Nir Eyal

1.

Joshua Burwell, age thirty-three, was visiting San Diego in December of 2015 from his home in Sheridan, Indiana. As he strolled along scenic Sunset Cliffs after Christmas dinner, he plunged sixty feet to his death. Bill Bender, a San Diego lifeguard, told reporters, "Witnesses stated seeing someone distracted by an electronic device and he just fell over the edge. [He] wasn't watching where he was walking[;] he was looking down at the device in his hands." While the death of any screen-addicted person is tragic, what is also worrying is that we are building whole sectors of the digital economy on the concept of addiction. We check our phones about 150 times a day — an average of every 6.4 minutes — according to a 2013 report.

From 1990 to 1991, I produced Wim Wenders's epic road movie, *Until the End of the World,* set in 1999 on the eve of the millennium. One of the plot themes was the addiction many characters had to the small handheld screens they carried around. As a critic for the *Guardian* wrote, "What begins as a natural desire to understand one's own past becomes an addiction to nostalgia. In 2015, their zombie-like wanderings, as they clutch handheld screens, take on new meaning. Of all the predictions Wenders made, this is by far the most striking."

What does it mean to be human in the age of digital addiction? Or, more pointedly, if so much of our humanity is expressed online, what do we make of the fact that our mode of self-expression is the raw material on which a handful of companies grow rich?

On November 13 and 14, 2015, around one thousand activists and techies met at the New School in New York City to talk about reinventing the Internet. Their hope was to create a co-op model: individuals dealing directly with one another without having to go through data-scraping corporate hubs such as Google and Facebook. Douglas Rushkoff, a social critic and writer, noted that there's too much focus on making sure that new innovations will be good for the *machines.* "I'm on Team Human!" he said at the conclusion of his talk.

The problem is that Internet designers are not treating us like humans; they're treating us like lab rats. Shortly after it was published I read a book called *Hooked: How to Build Habit-Forming Products.* The basic thesis of this very successful book is that in order to gain admission to the digital winner's circle you need to get your customers addicted to your app. The "trigger, action, reward, investment" sequence is curiously close to that of the Skinner box we all studied in Psychology 101. As author Nir Eyal explains it, "At the heart of the Hook Model is a powerful cognitive quirk described by B. F. Skinner in the 1950s, called a variable schedule of rewards. Skinner observed that lab mice responded most voraciously to random rewards. The mice would press a lever and sometimes they'd get a small treat, other times a large treat, and other times nothing at all. Unlike the mice that received the same treat every time, the mice that received variable rewards seemed to press the lever compulsively."

Like the poor lab rat in pursuit of happiness who clicks on a bar for a reward pellet, we spend hours looking for "like" rewards on our social networks. Those with the most likes turn it into a form of currency, as was demonstrated in the myriad "gifting suites" at the 2015 Sundance Film Festival, where popular YouTubers like iJustine "shopped" for free merchandise in exchange for posting about the swag to her three million subscribers. iJustine, whose fame stems from having posted a video on YouTube about her three-hundred-page iPhone bill, noted to the *New York Times,* "I love products, and I love sharing if I love something. Like, you can probably guarantee that it's going to be posted, especially if I love it."

2.

It would be easy to diss iJustine's blurring of the line between opinion and eager receipt of free products if it weren't the basic currency of the Internet age. What is the Kardashian empire but a walking product-placement opportunity? How would the TV and movie business survive without "brand-integration" dollars to top off the budget? And how would *Vox*, *BuzzFeed*, or even the vaunted *Atlantic* survive without the native advertising that blurs the line between editorial content and paid advertising? If indeed the author of *Hooked* is onto something – exposing our powerful addictions to social networking apps – then is Peter Thiel's almost spiritual commitment to "liberty" really the same as Thomas Jefferson's life, liberty, and the pursuit of happiness? I don't think so. Is your friend who spends three hours a day on Snapchat really free? What agency do any of us have in our relationship with Facebook?

I am just as guilty as you are. I surrender all my personal data to Facebook in return for the ability to share my vacation photos with my friends. But Twitter and Facebook want more than the personal data of any one user. Each platform has become a critical tool in political campaigns. During the 2016 presidential campaign, Donald Trump regularly boasted about his ten million Twitter followers, even though (according to the site StatusPeople, which tracks how many Twitter accounts are bots, how many are inactive, and how many are real) only 21 percent of Trump's Twitter followers are real, active users on the platform. Hillary Clinton didn't fare much better, with only 30 percent of her followers classified as real.

During the 2012 presidential race, the Annenberg Innovation Lab studied Twitter and politics, and what we found was pretty disturbing. We created a natural-language-processing computer model that read every tweet about every candidate and sorted them by sentiment. At the beginning I loved reading the dashboard of the twenty most positive and negative tweets of the previous hour. But within weeks the incredible amount of racist tweets directed at our president became too painful to look at. The anonymity that Twitter provides is a shield that brings out the worst in humans. Plato (*Republic* 2.359a–2.360d) told a tale of the Ring of Gyges: when put on, it would render the wearer invisible. He asked the question: If we were shielded from the consequences of our actions, how would that change the way we act? We know the answer.

As David Brooks says, we have created a "coliseum culture" in which a celebrity gets thrown to the lions on a weekly basis. Punishing strangers ought to be a risky endeavor. They can strike back and therefore threaten our long-term survival. Evolution, as Darwin pictured it, favors narrow self-interest. But the anonymity of the Internet shields the person who punishes the stranger. But also rewards us for exaggerating how we feel — the more outrageous tweet gets noticed. "It seems like our brains are wired to enjoy punishing others," the psychologist Nichola Raihani said to *Wired.*

Andy Warhol foretold our present YouTube lifestyle when he said, "In the future everyone will be world-famous for fifteen minutes." It was said with a touch of irony, because it was really a response to the question of why he would make a fairly homely drag queen like Holly Woodlawn one of his superstars. The brilliance, of course, lies in Warhol's qualifier "*fifteen minutes.*" The fact that twenty-one reality TV stars have committed suicide between 2005 and 2015 confirms Warhol's vision of the fleeting nature of fame in the twenty-first century.

3.

Perhaps the Internet's greatest failing is the convergence of anonymity and misogyny. When Peter Thiel noted that giving the vote to women in the 1920s "rendered the notion of 'capitalist democracy' into an oxymoron," he was reiterating a basic Randian trope — men are makers and women are takers. In tech employment, women make up 29 percent of the workforce even though they comprise 47 percent of the US labor force, according to the US Census Bureau. Even more troubling is that women led only 8 percent of all start-ups funded by venture capitalists in 2015. But what is most disturbing are the attitudes toward women that get displayed in an incident like Gamergate.

Zoe Quinn was an oddity, a twenty-eight-year-old female video-game developer. Her life was good until she split with her boyfriend, as the *Washington Post* tells it:

> Since August 2014, when her ex-boyfriend, Eron Gjoni, posted a 9,000-word screed about her online, Quinn and her

> family had been deluged by threats so severe that Quinn fled her home in Boston, afraid for her life. Gjoni's online "hate mob," as Quinn described it to a municipal judge that September, had unearthed her address and old nude photos, hacked her website, promised to kill and rape her, and placed multiple threatening calls to her father's home in upstate New York – all of which Quinn had meticulously documented and organized in evidentiary zip-drive folders.

But the problem wasn't just the scorned lover. It was the Internet mob of young men with a deep hatred of women and their efforts to be part of the gaming community. Because other women came to Zoe's defense, the mob has "morphed into a neo-reactionary movement, bent largely on fighting 'social justice warriors' online." But for the young men spending hours harassing Zoe Quinn, it is feminism that is the problem. Of course Peter Thiel felt this back in the early 1990s when he wrote in the *Stanford Review* that "the passionate hatred of men, the utopian demands for an elimination of all gender differences and the belief in widespread gender discrimination" were the biggest problems on the Stanford campus. He felt that this political correctness had to end.

The next generation of radical libertarians has called itself the alternative right, though many have noted this is just a less offensive title than "fascist." Milo Yiannopoulos, the Breitbart News Network columnist who was Eron Gjoni's biggest defender, described a characteristic of the group: "The alt-right's intellectuals would also argue that culture is inseparable from race. The alt-right believe that some degree of separation between peoples is necessary for a culture to be preserved. A Mosque next to an English street full of houses bearing the flag of St. George, according to alt-righters, is neither an English street nor a Muslim street – separation is necessary for distinctiveness." Asked why he supported Trump, Milo replied, "Trump represents the best hope we have of smashing political correctness apart. . . . I want a new political re-alignment on libertarian-authoritarian lines rather than left and right." And when Trump hired the CEO of Breitbart, Steve Bannon, to run his campaign, a *Washington Post* columnist noted that it represented "the dangerous seizure of the conservative movement by the Alt-Right." But for Milo, Peter Thiel is a hero on par with Trump for funding the Hulk Hogan lawsuit against *Gawker.* "With his lawsuit," he wrote, "Thiel has perhaps done more than any man to liberate social media from the terror of left-wing public shaming that prevailed in the golden age of Gawker. Indeed, nothing underscores the end of the reign of Gawker more than the rise of the alternative right, who cannot even be shamed by conservative media."

The distance from Thiel to the alt-right is not long, but the terms of engagement have changed. At first Thiel signed his opinions proudly (both in the *Stanford Review* and in his books), but then, like most alt-right posters (with the exception of Milo), Thiel tried to hide his involvement through anonymity – as did the young man who wrote Zoe Quinn, "I'm not going to stop spreading your disgusting nudes around and making sure your life is a living hell until you either kill yourself or I rape you to death."

When the police actually found some of the worst of Zoe's harassers, they couldn't arrest them because they were only thirteen years old.

4.

To counter the meme that social networks lead to antisocial behavior, the PR wizards at Facebook and Google pushed forth the notion that social networks are the founts of democracy. In May of 2011, on the heels of the Arab Spring demonstrations, Google's Eric Schmidt and Facebook's Mark Zuckerberg each put on a suit and tie to appear at the G8 summit in Deauville, France. They were there to persuade world leaders to eliminate any regulation on their companies' ability to collect data on their users. Zuckerberg pleaded, "People tell me on the one hand, 'It's great you played such a big role in the Arab spring, but it's also kind of scary because you enable all this sharing and collect information on people.' But it's hard to have one without the other. You can't isolate some things you like about the internet, and control other things you don't." So his pitch was, "If you want me to continue bringing down dictators, don't regulate my business." But a little distance from the Arab Spring lets us see that this is really a false choice.

Wael Ghonim, the Egyptian Google employee who helped launch the Tahrir Square revolution in early 2011 that toppled Hosni Mubarak, tells the real story. His words come from a TED talk he gave after he was freed from jail and escaped Egypt.

I once said, "If you want to liberate a society, all you need is the Internet." I was wrong. I said those words back in 2011, when a Facebook page I anonymously created helped spark the Egyptian revolution. The Arab Spring revealed social media's greatest potential, but it also exposed its greatest shortcomings. The same tool that united us to topple dictators eventually tore us apart.

So what went wrong? According to Ghonim:

First, we don't know how to deal with rumors. Rumors that confirm people's biases are now believed and spread among millions of people.

Second,... we tend to only communicate with people that we agree with, and thanks to social media, we can mute, un-follow, and block everybody else.

Third, online discussions quickly descend into angry mobs.... It's as if we forget that the people behind screens are actually real people and not just avatars.

And fourth, it became really hard to change our opinions. Because of the speed and brevity of social media, we are forced to jump to conclusions and write sharp opinions in 140 characters about complex world affairs. And once we do that, it lives forever on the Internet....

Fifth — and in my point of view, this is the most critical — today, our social media experiences are designed in a way that favors broadcasting over engagements, posts over discussions, shallow comments over deep conversations. It's as if we agreed that we are here to talk at each other instead of talking with each other.

Though Ghonim was talking about his post-Tahrir experience, his is an almost eerie description of political conversations on Twitter in the United States and elsewhere. Social media may have been useful in getting people to come to a demonstration, but it proved useless in trying to organize an opposition once Mubarak had fallen. Perhaps more troubling is the fact that, according to the nonprofit organization Freedom House, autocracy has grown across the world since 2011, with seventy-two countries experiencing a decline in freedom and only forty-three improving. The organization notes the ability of governments such as those of China, Saudi Arabia, and Syria, to use the Internet as a tool for spying on their populations and points out that government monitoring of social media has increased dramatically since 2011. As Chinese blogger Su Yutong noted after she was forced to flee her country, "Some people say that for China [the] Internet is a gift from God. However, for Internet users in China, it is more like dancing in shackles." But the Chinese plan to go much further in using the Internet to monitor their people, as the *Washington Post* reported in October of 2016:

> Imagine a world where an authoritarian government monitors everything you do, amasses huge amounts of data on almost every interaction you make, and awards you a single score that measures how "trustworthy" you are.
>
> In this world, anything from defaulting on a loan to criticizing the ruling party, from running a red light to failing to care for your parents properly, could cause you to lose points.

And in this world, your score becomes the ultimate truth of who you are — determining whether you can borrow money, get your children into the best schools or travel abroad; whether you get a room in a fancy hotel, a seat in a top restaurant — or even just get a date.

This is not the dystopian superstate of Steven Spielberg's "Minority Report," in which all-knowing police stop crime before it happens. But it could be China by 2020.

It is the scenario contained in China's ambitious plans to develop a far-reaching social credit system, a plan that the Communist Party hopes will build a culture of "sincerity" and a "harmonious socialist society" where "keeping trust is glorious."

5.

When tech plutocrats invoke the concept of freedom, what do they mean? Dr. Martin Luther King Jr. led the March on Washington for Jobs and Freedom. It's obvious now that the freedom brought to us by the libertarian elite will not come with jobs. The fact that Facebook is on track to generate annual revenues of $20 billion with fewer than fifteen thousand employees speaks volumes. Is Peter Thiel's idea of corporations — *free* to reap monopoly profits and operate *free* from government regulation — what we want for our country? Thiel's icon Ayn Rand defined freedom this way: "To ask nothing. To expect nothing. To depend on nothing." How far is this from Jefferson's great inspiration, the Greek philosopher Epicurus, who defined the good life and freedom in the following terms?

- The company of good friends.
- The freedom and autonomy to enjoy meaningful work.
- The willingness to live an examined life with a core faith or philosophy.

If we think about the world techno-utopians are envisioning, it may be hard for the average citizen to have the freedom and autonomy to enjoy meaningful work. Would a life where your daily existence relied on driving four hours a day for Uber, serving as a concierge for your Airbnb guests in the spare room, and spending your evenings doing crowdwork on Amazon's Mechanical Turk meet Epicurus's test? And would you have any time to live an "examined" life? Is the goal of tech success freedom, or addiction?

6.

The New Camaldoli Hermitage perches 1,300 feet above the Pacific in Big Sur, California, and is void of cellular service, Wi-Fi, and all other electronic conveniences. For my sixty-eighth birthday, inspired by a little book by Pico Iyer, *The Art of Stillness: Adventures in Going Nowhere,* I gave myself the gift of time and peace at this Benedictine retreat. Aside from the chanting of the monks in the chapel, no words are spoken there — which is of course the point. At home I am just as guilty as anyone of trying to watch TV, check my email, and talk on the phone at the same time. But I think we all have to take vacations from our devices.

In my three days with the monks, I learned that the Benedictines believe that life revolves around five practices.

- Prayer: This can be any daily silent practice or meditation.
- Work: This becomes part of a balanced life. It cannot be the whole focus.
- Study: This is when we read the wisdom of those who came before us.
- Hospitality: This just means treating those around you with kindness; it can also mean breaking bread with friends.
- Renewal: This is accomplished by taking one day a week to turn away from daily cares (and screens) and appreciate the natural beauty around us.

I am not Catholic, yet I find the monks' prescriptions to be helpful, a model of how I want to live in the world. The idea of an examined life is missing in our current digital rush. Perhaps following the monks' example of devotion to their community would be too much of a sacrifice for most of us, but when I was immersed in their fourteenth-century songs my mind kept wandering to the events that had occurred two weeks before in Charleston, South Carolina, where nine church parishioners were killed by a racist kid named Dylann Roof. When you think that the families of the slain churchgoers were able to forgive the shooter, you can only marvel at the power of their faith. Never was the difference between community cooperation and individual separation more starkly outlined. I'm not sure my faith would afford me that amount of grace in the face of such evil, but I am awed to see it exist in this hateful political climate we inhabit. I kept thinking of how powerful this sense of community was.

The great biologist E. O. Wilson makes the argument in *The Social Conquest of Earth* that evolution favors humans who learn how to cooperate. Some of our prehistoric ancestors went out to hunt, while others stayed and kept the fire going. If everyone went out to hunt for his or her own food, there would be no fire over which to cook it when the hunters got back. This is why I have no truck with the arguments of Peter Thiel and his fellow Ayn Rand acolytes, who believe that "if any civilization is to survive, it is the morality of altruism that men have to reject." Right now we are inundated with Silicon Valley propaganda about surviving in "the shark tank" — we idolize the take-no-prisoners entrepreneur who will stop at nothing to build his company. The biologist Frans de Waal describes how Enron CEO Jeffrey Skilling destroyed his own company.

> An avowed admirer of Richard Dawkins' gene-centric view of evolution, Skilling mimicked natural selection by ranking his employees on a one-to-five scale representing

> the best (one) to the worst (five). Anyone with a ranking of five got axed, but not without first having been humiliated on a website featuring his or her portrait. Under this so-called "Rank & Yank" policy, people proved perfectly willing to slit one another's throats, resulting in a corporate atmosphere marked by appalling dishonesty within and ruthless exploitation outside the company.

Since at least 1995, professors at business schools have dismissed this kind of behavior as a natural outgrowth of the Austrian economist Joseph Schumpeter's notion of "creative destruction." The growth of the tech economy with its constant change would create a new kind of employee: oriented to the short term and focused on potential ability rather than acquired knowledge. But most of us are like Epicurus or even the monks of Camaldoli. We need a life narrative in which we take pride in being good at a specific task, and we value the experiences we have lived through.

Many of my colleagues on the faculty of USC certainly approach their work with this reverence for knowledge, but their students have new pressures on them to adopt the culture of creative destruction. They are expected to get a well-paying job the minute they leave college. No more thoughts of bumming around Europe for a year before you "get serious." And the reason for this is the specter of debt that hangs over the head of most middle-class college kids. In 1970 the University of California raised the tuition for in-state students to $150 a year. Today the in-state tuition is $13,500. The effect of the average student debt of $30,000 upon graduation is to increase the pressure to get a good job. As economist Joseph Stiglitz has written, "On average, many college graduates will search for months before they find a job – often only after having taken one or two unpaid internships. And they count themselves lucky, because they know that their poorer counterparts, some of whom did better in school, cannot afford to spend a year or two without income, and do not have the connections to get an internship in the first place." And at least from the vantage point of a California college, getting a job seems to mean work in the technology sector. Back East, that probably means a job on Wall Street. In essence our universities are being turned into trade schools. Are we abandoning the humanities and a basic liberal arts education for the sake of preparing students for the careers that await them in Silicon Valley or on Wall Street?

The liberal arts and the humanities matter because, as the Greeks in Epicurus's time would say, without art we have no empathy. The ancient Greeks used the word *catharsis* to mean the purification of emotions through art that results in renewal and restoration. The cultural critic Leon Wieseltier notes, "By creating sympathy, art lays the ground, the internal condition, for moral behavior." I've been lucky enough to experience these feelings of catharsis many times – in concerts by Bruce Springsteen and Bob Dylan; in movies by Stanley Kubrick, David Lean, and Martin Scorsese. I felt it reading *To Kill a Mockingbird* as a teenager. My guess is that you, too, have felt the emotional renewal of art on many occasions. But there are periods (perhaps the present one) when the desire to spark catharsis slips into the background and performers find themselves faced with more commercial pressures.

In his seminal 1976 work, *The Cultural Contradictions of Capitalism,* Harvard sociologist Daniel Bell contended that modern capitalism creates a culture of such self-gratification and narcissism that it may end up causing its own destruction. This idea seems like a perfect way to think about DigiTour. DigiTour is a sixty-city concert tour of the six most popular YouTube stars. In 2015, they sold more than 220,000 tickets, mostly to girls between the ages of nine and fifteen. At points the din from the screaming fans gets so loud that the security guards stuff Kleenex in their ears. This, of course, is an American ritual, perhaps starting with the appearance of Frank Sinatra at New York's Paramount Theatre in December of 1942. Jack Benny commented, "I thought the goddamned building was going to cave in. I never heard such a commotion.... All this for a fellow I never heard of." In 1956 Elvis Presley drew similar screaming crowds, and of course in 1964, Beatlemania swept across the country.

But at DigiTour, something was different — the "talent" didn't sing or dance at all. As *BuzzFeed* noted: "By and large, the cast do not really perform so much as appear. Roughly once every show, a booming voice prods, 'Now, let's — take — some — SELFIEEEES,' in the way another announcer might implore a crowd to make some noise. The fans oblige." The biggest star on YouTube is PewDiePie, who has more YouTube followers than Beyoncé. There are thirteen billion views of his videos, which feature him playing (and exaggeratingly reacting to) video games. That is the new talent. I'm sure it has nothing to do with catharsis or any of the roles art has played in our lives throughout history.

The Internet revolution was supposed to usher in a new age of digital democracy, opening up distribution pipelines to anyone with talent. But what are we to make of a teenage phenom whose sole talent is playing a video game? The famous infinite monkey theorem posits that if you let enough monkeys type for a long enough time they will eventually write *Hamlet*. But have the four hundred hours of video uploaded to YouTube *every minute* produced the new Scorsese or Coppola? Could it be that the economics of "more" is drowning us in a sea of mediocrity?

Of course there are those who hold on to the notion that we are living in a golden age of television. But it must be noted that all the great programs usually cited as proof — *The Sopranos, Breaking Bad, Mad Men, The Walking Dead, True Detective* — exhibit a dark tone of nihilism that may perfectly match what the cultural theorist Jacques Barzun calls our age of decadence: "The forms of art as of life seem exhausted, the stages of development have been run through. Institutions function painfully. Repetition and frustration are the inevitable result. Boredom and fatigue are great historical forces."

Now, my USC colleague Henry Jenkins, the great cultural critic, has another take on this, and I owe it to you to give you his opinion on the current state of TV:

> I do think [these dark dramas] may also have some sociological significance since what is being questioned here most

> persistently is a certain kind of white masculinity — which emerges here as in crisis. The power of old institutions seems to be in decline, certain kinds of privilege is being questioned, certain hypocrisies are no longer tolerated. These stories can be read by white men as speaking to their decline from power and from other groups as representing a critique of traditional power structures. The ambivalence associated with the anti-hero speaks in both ways, and thus, the dual address itself reflects the fragmentation and discontent at the center of the American dream today.

So this notion of an antihero who breaks all the rules and gets away with it (Don Draper, Tony Soprano) is dominant, but Jenkins notes that there is another, more hopeful set of TV genres, one that counters the decline of white male:

> So, we have stories of bright young women, especially women of color, trying to achieve success in their chosen fields, often modeling different kinds of support structures, often pushing back against structural racism, often dealing with their own self-doubts, as a contrast to the swaggering, but ultimately, flawed powerful white men. And then we have stories of teams that work together for the public good [police procedurals], often facing limited resources and moral ambiguities, but at heart, as a group, following a code of ethics and a rational process that arrives at sound answers.

Jenkins may be right that the golden age of television is real and sustainable, but I still think the winner-takes-all economics of other media, including music, film, books, and journalism, will eventually break the TV business. As the FX network CEO, John Landgraf, pointed out, "There is simply too much television. We're in the late stages of a bubble." I was lucky enough to have dinner with the late William Paley in the mid-1980s. Paley had founded CBS and run it for more than fifty years, and he told me that owning a broadcast TV station in the 1960s was "a license to print money." A top show could regularly get an audience of forty million viewers on a single night. In its final season, in 2015, *Mad Men* was lucky to get 1.7 million viewers. Who knows how many people watch *Mozart in the Jungle,* Amazon's Golden Globe–winning drama? My guess is it's less than a million. The production costs of a one-hour drama have never really fallen, even though the audience has declined by 80 percent. The law of diminishing returns in an era of proliferating production (especially from Amazon, YouTube Red, and Netflix) will eventually cause a crash.

But the wizards at Google, Facebook, and Amazon assure us that data will prevent the crash. Bloomberg, summarizing a report from JP Morgan Chase, wrote, "Whoever has the most data comes out on top, and the value of metadata will shift from content distribution to actual production." Perhaps, but if the music business is any guide, this is a dead end. The business has become a song machine, as John Seabrook points out in his book of the same title. Seabrook attributes the uniformity of pop tunes to the use of data, which pushes producers toward the "track and hook" songwriting method now in vogue, especially in hip-hop and dance music.

> In a track-and-hook song, the hook comes as soon as possible. Then the song "vamps" — progresses in three-or-four chord patterns with little or no variation. Because it is repetitive, the vamp requires more hooks: intro, verse, pre-chorus, chorus, and outro hooks. "It's not enough to have one hook anymore," Jay Brown explains. "You've got to have a hook in the intro, a hook in the pre, a hook in the chorus, and a hook in the bridge, too." The reason, he went on, is that people on average give a song seven seconds on the radio before they change the channel, and you got to hook them.

Now, if the major record labels were pouring more of the proceeds from the song machine into fostering new talent rather than making more factory-produced pop, we might find ourselves in a time of artistic renewal. Jacques Barzun, writing about the Renaissance, noted that prolific periods of creativity show up in history for relatively short periods of time and then disappear. So perhaps we are just in a creative interregnum — one in which the machine has taken over for the artist. And another renaissance will come again.

7.

I was lucky enough to live and work in the midst of a couple of renaissance moments in the 1960s music business with Bob Dylan and The Band and George Harrison. And then starting in 1974 I was a producer in Hollywood at a time when a revolutionary group of young filmmakers – including Martin Scorsese, Francis Ford Coppola, George Lucas, Steven Spielberg, and Paul Schrader – had both the "feverish interest" and the appetite for rivalry to remake the film business. From the point of view of old Hollywood, the lunatics had taken over the asylum, and an extraordinary creativity reigned. It all lasted around six years, then the suits took control again. But in the interim, what glorious films they were: *The Godfather, American Graffiti, MASH, Nashville, Taxi Driver, Jaws, Five Easy Pieces, Shampoo, The Last Picture Show, Mean Streets, The French Connection,* and *Badlands,* to name just a few. Those of us involved in this American New Wave will treasure those years forever. By 1980, the music and the film businesses had both migrated to the blockbuster model epitomized by Michael Jackson's *Thriller* and George Lucas's *Star Wars.* It was as if the entire lineup of a baseball team was expected to hit home runs instead of singles and doubles.

What made this extraordinary period possible? To begin with, it came on the heels of a crash. The movie business in the late 1960s was dominated by five family-owned firms that were in only one line of business – the production and distribution of movies. Warner Bros. was owned by Jack Warner, 20th Century Fox was controlled by the Zanuck family, and Columbia Pictures was run by the Schneider family. United Artists was owned by Arthur Krim, and Paramount was still controlled by the nonagenarian Adolph Zukor. As the TV business got more successful, movie attendance continued to drop. By the early

1960s the movie business was in crisis. In an attempt to grow their audience, the studios produced a series of spectacles and musicals. Although these films were not too different from the standard fare, they were much more expensive. Films like *Cleopatra, Paint Your Wagon, The Molly Maguires,* and others were huge flops, and by 1969 most of the studios were teetering on the edge of bankruptcy. In the midst of this chaos a young man named Bert Schneider persuaded his older brother Stanley, who ran Columbia Pictures, to give him $360,000 so he could make a hippie motorcycle road movie. Released in early 1969, *Easy Rider* completely changed the industry. The film grossed more than $60 million, so the return on investment was extraordinary. Almost instantly, every broke studio began looking for first-time directors who could make films for less than $1 million.

The obvious place to look was the film schools — NYU, USC, and UCLA. But the young men (and they were almost all men) who were training in these schools had a very different view of the role of the director from that of the old moguls who still ran the studios. The kids had been schooled on European New Wave directors such as François Truffaut, Jean-Luc Godard, Federico Fellini, and Ingmar Bergman. In Europe, the director was the auteur of the film, both legally and in spirit. After the success of *Easy Rider,* the film's director, Dennis Hopper, asked for and received final cut privilege — i.e., the right to decide what is and isn't included in the version of the film released to the public — and that became standard procedure at

Bert Schneider's company, BBS, which went on to make *Five Easy Pieces* with Bob Rafelson and *The Last Picture Show* with Peter Bogdanovich. All this required a new kind of executive who could be sensitive to the needs of this new kind of filmmaker. Arthur Penn, who made *Bonnie and Clyde* for Warner Bros., tells the story of an old Jack Warner trying to understand a new generation. Warren Beatty, the star and producer of the film, went with Penn to Jack Warner's house to screen it for him. Warner said, just before the lights went down, "If I have to get up and pee, you know the picture stinks." Twenty minutes in, Warner got up and peed. When it was all over, he turned to Penn and Beatty and said, "What the hell is this? Where are the good guys?" He hated the picture, but having sunk $2.5 million into it, he allowed Beatty to roll it out in what was known as an art-house release. This meant opening it in a few theaters in key cities and hoping the critics and word of mouth would get the film noticed. Penn and Beatty were able to get strong support from the critics, and Warner's $2.5 million investment turned into a $70 million box-office return.

This brief respite from the economic doldrums allowed Warner to sell out to Seven Arts and then to Steve Ross, the owner of Kinney National Services, a parking-lot and janitorial business. But Ross was smart and realized that he needed a new breed of manager to run a creative business. He hired the talent agent Ted Ashley and the producer John Calley to run Warner Bros. Pictures. In many ways Calley followed the model of Mo Ostin, who ran

Warner Bros. Records (see page 39) for Ross. The record guys called it A and R — artists and repertoire. That's what you had to pay attention to. Keep the artists happy, and make sure the songs and scripts (repertoire) are of the highest quality. For example, the first half hour of the meeting Martin Scorsese and I had with John Calley after he bought our movie *Mean Streets* was spent talking about Miles Davis, a mutual enthusiasm for the three of us. Calley's office furniture consisted of two long suede sofas surrounding an Eames chair and ottoman, where Calley sat — not a desk in sight. The idea was to break down the hierarchy, and it worked. We left Calley's office feeling like we had been admitted to a secret society.

Artist-friendly management was only part of the deal. The secret sauce was frugality. Most of the movies I cited in connection with this renaissance were made for around $1 million. Even the epic film of the era, *The Godfather,* only cost $6 million. Given that *Cleopatra,* filmed ten years earlier, had cost $37 million, *The Godfather* was a bargain for Paramount. The trade-off was artistic freedom in return for budgetary restraint. Very few directors have final cut privilege anymore, because the very idea of asking a studio to give you $100 million and then telling the suits they have no say in the film is beyond the pale, even for the most inflated Hollywood egos. And because budgets were smaller and artists had more freedom in the 1970s, the films inevitably traversed into the realms of politics, sex, and power as opposed to the power of a superhero in a fantasy universe. This was of course quite edifying for a generation who came of age with the Pill and the assassinations of John F. Kennedy, Dr. Martin Luther King Jr., and Robert Kennedy, not to mention the Vietnam War. The artists working on a low budget on the margins of the culture are usually those who advance their art. Orson Welles's budget for *Citizen Kane* was $800,000, just two years after MGM had spent nearly $4 million on *Gone with the Wind*. It is universally acknowledged that *Kane* advanced the art of cinema and was the greater picture. The big studios have pretty much abandoned those marginal artists and their aspirations and now concentrate their capital and their minds on "franchises" – endlessly duplicatable comic books that can be built to a formula and require a traffic cop rather than an auteur.

What was most remarkable about that brief period from 1969 to 1979 was the collegiality we enjoyed and shared with one another. The film-school generation read each other's scripts, made casting suggestions, and sat in on rough-cut screenings. Of course there was rivalry, but – more important – there was also support. Francis Coppola, having seen an early cut of *Mean Streets*, suggested to Bernardo Bertolucci that he cast Robert De Niro in *1900*, told Ellen Burstyn to hire Martin Scorsese for *Alice Doesn't Live Here Anymore*, then cast De Niro in *The Godfather: Part II*. Paul Schrader came to an early *Mean Streets* screening and immediately called Michael Phillips, who had just won an Oscar for *The Sting*, and told him that Scorsese and De Niro would be perfect for his new script, *Taxi Driver*.

Could it be that the extraordinary period in America between the arrival of Louis Armstrong and Charlie Chaplin in the early 1920s and the advent of the blockbuster in

1980 represents an unusually long renaissance that may not come this way again? In 1998, when the American Film Institute first published its list of the one hundred greatest films of all time, only thirteen of them were made after 1980. I'm not interested in nostalgia but rather in figuring out what changed. Perhaps Henry Jenkins is right: the neorealism that was so much a part of the American New Wave in the 1970s has drifted into TV. But TV has also spawned an age of reality shows in which Kim Kardashian and Donald Trump can overwhelm any cultural innovation that might exist.

8.

In 1970 the Nobel Prize–winning economist George Akerlof published a paper that may help us understand the effect that the commoditization of media by Facebook, YouTube, and Google is having on our culture. The paper was called "The Market for 'Lemons': Quality Uncertainty and the Market Mechanism." Akerlof says that when you buy a used car you assume the worst – it's a lemon – in your negotiation stance. Thus the seller of a really good used car always loses out. No one will pay for more than average quality. The typical consumer of ad-supported media in our broadband universe is like that used-car buyer: he or she assumes that content is of average quality, and this inevitably gives rise to what Chris Anderson, in his book *Free: The Future of a Radical Price*, claims is a business strategy essential to companies' survival – giving things away. If the YouTube video I'm about to watch is free, I lose nothing except my time watching it. I can assume it is a lemon, or at the very least worthless. In a world where YouTube, Google, and Facebook treat all content as a commodity – a means to an end – it isn't surprising that it's devalued in the public mind. The false democracy that places art and random uploads side by side has, I believe, led too many to believe that art is easy to make and therefore not valuable. As James DeLong said, "Google can continue to do well even if it leaves providers of its [content] gasping like fish on a beach."

Once upon a time a critic such as Pauline Kael or Greil Marcus could help me navigate this information asymmetry, driving interest in work they deemed to be of quality. The hegemony of the blockbuster in music, books, and film allows marketers to bypass critics, rendering them almost irrelevant to the process. Do you think the producer of Michael Bay's *Transformers* spends one minute worrying about critics? And don't confuse the critics I'm talking about with the crowd at Rotten Tomatoes. The great critic Pauline Kael was someone I trusted deeply. She was not an algorithm, processing twenty thousand comments about a film. She made an argument concerning each film – passionately reasoned – whether she liked it or hated it. Each work of art had a reason for being, and in championing the good stuff, she made us all better filmmakers. By contrast, Leon Wieseltier asserted that current criticism is all about "takes": "Make a smart remark and move on. A take is an opinion that has no aspiration to a belief, an impression that never hardens into a position."

The Internet fostered the take. A whole movie reviewed in a single tweet. It may be quite easy to review the latest *Transformers* movie in 140 characters, but I doubt that would work for *The Big Short* or *Spotlight*. The rationale behind a comic-book movie or even a new PewDiePie YouTube episode is that a tweet or a thirty-second ad on the Internet will tell you all you need to know about it.

So in the face of the commoditization of content on platforms like YouTube, I want to reassert that art is important. There is a sense in which art, politics, and economics are all inextricably and symbiotically tied together, but history has proved to us that art serves as a powerful corrective against the missteps of the establishment. There is a system of checks and balances in which, even though the arts may rely on the social structures afforded by strong economic and political systems, artists can inspire a culture to move forward, reject the evils of greed and prejudice, and reconnect to its human roots. This ability cannot be taught to a computer algorithm, despite what Veritas, a company that sells "information management solutions," claims in a piece of "sponsored content" in the *New York Times*:

> Lynn [Joshua Lynn of Piedmont Media Research] and his team are hired by major studios to test out new movie concepts with sample audiences. They start by tabulating hundreds of reactions to a written movie concept that Lynn says is "almost a textual description of what we might get in a trailer." The reactions are then analyzed to produce what the company calls a Consumer Engagement

> Score – a number from one to 1,000 that shows how well audiences jibe with a concept, broken down by age, race, geographical location and other factors. Lynn says the score can predict a Hollywood hit with 89 percent accuracy.

This is of course total nonsense, but it is perhaps indicative of why our movie screens are flooded with look-alike comic-book-hero movies. You can call it what you want, but it is no longer the art of the cinema.

Those seeking political and economic change should consider embracing art. Part of our role as citizens is to look more closely at the media surrounding us and think critically about its effects – specifically, whose agenda is being promoted and whether it's the agenda that will serve us best. In 2011, the screenwriter Charlie Kaufman (*Being John Malkovich, Adaptation*) gave a lecture at the British Academy of Film and Television Arts. He said something both simple and profound:

> People all over the world spend countless hours of their lives every week being fed entertainment in the form of movies, TV shows, newspapers, YouTube videos, and the Internet. And it's ludicrous to believe that this stuff doesn't alter our brains.
>
> It's also equally ludicrous to believe that – at the very least – this mass distraction and manipulation is not convenient for the people who are in charge. People are starving. They may not know it because they're being fed mass-produced garbage. The packaging is colorful and loud, but it's produced in the same factories that make

Pop-Tarts and iPads by people sitting around thinking, "What can we do to get people to buy more of these?"

And they're very good at their jobs. But that's what it is you're getting, because that's what they're making. They're selling you something. And the world is built on this now. Politics and government are built on this; corporations are built on this. Interpersonal relationships are built on this. And we're starving, all of us, and we're killing each other, and we're hating each other, and we're calling each other liars and evil because it's all become marketing and we want to win because we're lonely and empty and scared and we're led to believe winning will change all that. But there is no winning.

In his book *Amusing Ourselves to Death: Public Discourse in the Age of Show Business,* Neil Postman compares two midcentury views of the future: George Orwell's in *1984* and Aldous Huxley's in *Brave New World.* Though Orwell's dystopian nightmare is probably better known, Huxley's vision, in my opinion, was more true to our current moment. Following is Postman's beautiful summary:

> What Orwell feared were those who would ban books. What Huxley feared was that there would be no reason to ban a book, for there would be no one who wanted to read one. Orwell feared those who would deprive us of information. Huxley feared those who would give us so much that we would be reduced to passivity and egoism. Orwell feared that the truth would be concealed from us. Huxley feared the truth would be drowned in a sea of irrelevance.

> Orwell feared we would become a captive culture. Huxley feared we would become a trivial culture, preoccupied with some equivalent of the feelies, the orgy porgy, and the centrifugal bumblepuppy.

Huxley's assertion was that technology would lead to passivity. The ease with which we could consume mind-numbing entertainments and distractions would ultimately rot our democracy. And this is exactly what may be happening. In the 2016 presidential election in the United States, ninety-eight million citizens who were eligible to vote declined to exercise that privilege, according to the United States Elections Project. And a much larger percentage of millennials are nonvoters. As Kevin Drum reported in *Mother Jones,* "In 1967 there was very little difference between the youngest and oldest voters. By 1987 a gap had opened up, and by 2014 that gap had become a chasm." As much as we can rail about the deleterious effects of the Koch brothers' billions on the political process, if half the American electorate thinks that voting is a waste of time, we need to understand democracy's failure. If most of those young nonvoters are on social media, then Facebook is at least partially to blame. In Huxley's world, the obsession with taking drugs, going to the "feelies" (his equivalent of IMAX movies), playing interactive games, and downloading porn filled the lives of the citizens. They had no time for politics or even for wondering why their horizons were so narrow. The kids attending DigiTour would fit right into the plot of *Brave New World.* The Internet's self-curated view from everywhere has the amazing ability to distract us in trivial pursuits, narrow our choices, and keep us safe in a balkanized suburb of our own taste. Search engines and recommendation engines constantly favor the most popular options and constantly make our discovery more limited.

I began this chapter wondering whether technology was robbing us of some of our essential humanity. Google's chief technologist proclaims that technology will "allow us to transcend these limitations of our biological bodies and brains....There will be no distinction, post-Singularity, between human and machine." But the brilliant essayist Mark Greif notes, "Anytime your inquiries lead you to say, 'At this moment we must ask and decide *who we fundamentally are,* our solution and salvation must lie in a new picture of ourselves and humanity, this is our profound responsibility and a new opportunity' — just stop."

That is exactly what I propose: let's stop and consider a strategy of resistance to techno-determinism. Any historian will tell you that revolutions often overshoot their marks, a statement the citizens of Paris in 1789 or Moscow in 1925 would have agreed with. The digital revolution is no exception.

CHAPTER TWELVE

The Digital Renaissance

He not busy being born is busy dying.
— Bob Dylan

1.

If the Internet started out both decentralized and democratic, why can't we return to that state? I am under no illusion that this would be an easy process or that I have the correct strategy. I am not going to try to address the larger issue of whether robots and artificial intelligence are going to lead to a world without jobs, for that would take a book in itself. I have suggested that policy makers begin exploring a universal basic income, or UBI, a concept that has support on both the left and right. It does seem to me that to ignore the dystopian possibility that software will "eat the world" would be foolhardy. Just because some techno-optimists continue to insist that old jobs will be replaced by new jobs we can't imagine yet does not mean it is true.

Google's AlphaGo artificial intelligence system may have bested the world's greatest Go player, but I'm not worried that it's going to replace our greatest musicians, filmmakers, and authors, even though an NYU artificial intelligence laboratory has programmed a robot named Benjamin to be a screenwriter. And even if you believe that robots will be able to fill most jobs, MIT's Andrew McAfee and Erik Brynjolfsson have pointed out that "understanding and addressing the societal challenges brought on by rapid technological progress remain tasks that no machine can do for us." When I ask myself what it means to be human, I think that having empathy and the ability to tell stories rank high, and I am not worried that those skills will be replaced by AI. A great artist's ability to inspire people — especially to compel them to think and act — lies at the heart of political and cultural change. It really is the reason we tell ourselves stories. Plato teaches us that we should not expect the arts and humanities to be driven or dominated by the objectives of science. In fact part of the liberal arts' enduring mission is precisely to critique those objectives.

At the very beginning of this book, I wrote about computer scientists such as Alan Kay, who collaborated with many artists to make the Xerox Alto a tool that would "empower the artistic individual." But somehow, in the years that followed, artists took a backseat to coders. So we will need the technology and creative communities to work together again to reengineer the Internet. If we return to the original vision of the World Wide Web as described by Tim Berners-Lee, I think we will be on the right track. In 2014, in the face of the extraordinary consolidation of Internet power, Berners-Lee noted his concern about the "near-monopoly status" of Google and Facebook. His hope was that "by continually 're-decentralizing' the web, we will unleash the next generation of technology, business and social innovators." This requires both legal and business innovation, and it is those innovations to which we turn our attention.

2.

In August of 2001, at the Federal Reserve Bank's annual Jackson Hole, Wyoming, symposium, former treasury secretary Lawrence Summers and University of California economist J. Bradford DeLong delivered a paper on economic policy for the information economy. They began with a basic fact of digital economics: given that price equals marginal cost, "with information goods the social and marginal cost of distribution is close to zero." *Marginal cost* means the cost of producing one more unit of a good. Once a song is on a Spotify server, the cost of selling one more stream is zero. But here is the paradox:

> If information goods are to be distributed at their marginal cost of production — zero — they cannot be created and produced by entrepreneurial firms that use revenues obtained from sales to consumers to cover their [fixed set-up] costs. If information goods are to be created and produced... [companies] must be able to anticipate selling their products at a profit to someone.

In this zero-marginal-cost economy, the only way to make money is to scrape consumer data from your users and sell it to advertisers. In the creative world, nowhere does the fixed cost to produce high-quality music, video, books, and games get factored into this equation. How are musicians, journalists, photographers, and filmmakers going to survive in the zero-marginal-cost economy? For the media economy to continue, we are going to have to find ways to deal with the paradox that Summers and DeLong point to.

If zero marginal cost is the first economic fact, then the "law of large numbers" is the second one. At the Annenberg Innovation Lab, we believe that there will be at least five billion Internet-connected mobile devices in the world by 2018. If that's true, and if you could sell a streaming movie for $1 to 5 percent of that market, it would mean revenue of $250 million for that movie. The law of large numbers.

To see how that plays out in the rest of the market, try a little thought experiment with me. In 1998, the average yearly entertainment expenditure for movie tickets, event admissions, and other entertainment content costs for someone under twenty-five years of age in the United States was $393 (adjusted for inflation), according to the Bureau of Labor Statistics. By 2012, that amount had declined to $249 per year because of all the free entertainment available from YouTube, pirate sites, and social media. So assume that the richest 10 percent of those five billion mobile users will spend what a teenager did in 1998 — $28 per month. That would yield $170 billion in media revenue per year. Then assume the next 40 percent of mobile customers spend $10 per month — essentially a Netflix account. That would yield $240 billion a year. Finally, assume that the poorest 40 percent of users get their slightly more limited content for free in return for watching ads worth $5 per month. That would yield an additional $120 billion per year. This would mean a media and entertainment economy *just on mobile devices* of $556 billion per year. Given that the global media and entertainment economy is only $1.3 trillion from all sources, you can see that the future would be bright if we could only figure out how to make sure the creators of content get paid for this bounty.

For the media economy, the implications of zero marginal cost and the law of large numbers — in addition to the ideas I have put forth about advertising, peak content, and winner takes all — are profound. I could imagine a TV system with no more than forty "linear" channels, most of which would consist of live sports and news. The remaining bandwidth would be devoted to a huge selection of on-demand Internet video content, equipped with efficient search and preference tools. Almost all of what we now consider TV would be delivered on demand. This would mean that the Internet monopolists could also control TV. But it also raises the larger question of how Internet video content would be paid for in a world where advertising funds Google and Facebook. Kalkis Research has a daunting view of the future of online advertising:

> The online advertising market is saturated, and has no more room to grow. The traditional space for ads is overcrowded, and has started to shrink, as Internet users start to use ad blockers.

> Controls and regulations are nonexistent, and a big chunk of ad spending is being stolen, plain and simple. Customers are growing aware of the phenomenon of ad fraud. Every new fraud scandal bears the risk of customers scaling back on online ad spending. The whole ecosystem is at risk of turning from growth to decline, overnight, in a rerun of what happened in 2000–2001.

If indeed two hundred million people currently use ad blockers and the distinction between TV and the Web is dissolving, then we need to rethink how content gets paid for. Tim Berners-Lee, the Web's inventor, has an idea. "Ad revenue is the only model for too many people on the web now," Berners-Lee told the *New York Times*. "People assume today's consumer has to make a deal with a marketing machine to get stuff for 'free,' even if they're horrified by what happens with their data. Imagine a world where paying for things was easy on both sides." Berners-Lee speculates about a kind of micropayment system that would allow you easily to pay fifteen cents on your mobile bill for a piece of content. But until that is invented, my guess is that the most efficient model would be an all-you-can-watch subscription model that would not be too different from what you are paying now to your cable or telephone company. Essentially the carrier would bundle all these services across both fixed and mobile broadband. And because everything would be sent over the Internet, the cost of physical carriage and TV infrastructure would drop radically, so more money could flow to content creation. But to make that work, the current structure of Internet power must be reformed.

The basic laws governing the Internet were passed in 1998, at a time when it took four hours to download a movie and twenty minutes to download a song. Those laws may have made sense in an era before broadband, but today no individual can effectively police copyright infringement. We should begin with the changes to the law that are easy to make, then work our way up to the hard stuff. It makes sense to start with the music business — the first to get decimated by the digital revolution. In an interview, the great music manager Irving Azoff (the Eagles) stated the problem:

> It's very clear that the economics of streaming haven't helped creators. And I don't think that when President Clinton signed the law that led to safe harbor, the Digital Millennium Copyright Act, that there was any intention for the tech companies to hide behind it the way that they currently do.
>
> None of us in the music business have enough resources to fight the legal battles against these big tech companies. The best thing would be to change laws.

The DMCA "safe harbor" provisions that Google, YouTube, Twitter, and Facebook hide behind need to be changed, as Azoff suggests. Every artist who has tried to have his work removed from an infringing website knows the problem. The rightful owner of the work files a takedown notice, the infringing site deletes the file temporarily, then the user reposts the same work using a different URL. This game of cat-and-mouse keeps the material available somewhere. Under current US law, this procedure is perfectly legal, and it is used by every pirate site (as well as YouTube) to avoid prosecution. In the first twelve weeks of 2016, Google received takedown requests for 213 million links, representing a 125 percent increase over the same period in 2015. An amendment to the law needs to be written stating that once a takedown notice has been filed, it is the responsibility of the website to keep the content down (a system that some advocates are calling take down, stay down). Google complains that this would be technically hard to do, but we know the company has the automated tools that make it possible.

Google and YouTube are thinking ahead to a time when content creators may win the battle. They allow users to post anything on YouTube, citing the doctrine of fair use, which permits limited use of copyrighted material without the need to acquire permission from rights holders. Your local TV station can show a short highlight of a football game broadcast by another network under fair use. When you post something on YouTube, even if it is a whole song and even if you do not check the box that says "I own this piece of content," you check the box that says "fair use," and the content will go up. Google has made it very clear that it will increase the legal support it provides for its user communities so they can defend themselves against accusations of copyright infringement.

Academics and others who rely on quoting or otherwise using small portions of others' works are very aware that there are real limitations to fair use. You cannot use a work in its entirety and claim fair use. Musicians and filmmakers could rely on the Content ID system YouTube currently has in place, but few have the wherewithal to deal with every takedown counterclaim that is asserted on grounds of fair use. In a filing to the register of copyrights, Warner Music Group stated, "WMG estimates that currently, it would take at least 20–30 people, at a fully loaded cost in excess of $2 million per year, and probably the use of an outside content monitoring contactor at additional expense, to meaningfully affect (but not entirely block) just WMG's top 25 album releases on YouTube."

The solution to this stalemate is fairly simple. The Library of Congress should issue a set of guidelines to determine the precise definition of fair use. In academia, it usually means a video or audio clip cannot exceed thirty seconds and that it should be quoted in the context of some transformative work (such as a remix or mashup). The language USC uses is pretty clear:

> Did the unlicensed use "transform" the material taken from the copyrighted work by using it for a different purpose than the original, or did it just repeat the work for the same intent and value as the original?

If the Library of Congress were to issue guidelines on fair use, then when a YouTube user uploads a clip (asserting fair use) of a piece of work that Content ID identifies as being blocked by the copyright holder, the clip would be sent to a human screener employed by YouTube for evaluation — the same way YouTube screens out pornography and ISIS videos. If the clip does not meet the Library of Congress guidelines, it would stay blocked.

The second area of administrative law that we need to deal with concerns the Federal Trade Commission, which is in charge of regulating Internet advertising and data collection. Because Google and Facebook are mainly in the business of advertising and data collection, the FTC is really the most important regulatory agency. In the past, when the FTC has examined Google's advertising business, it has rejected the charge brought by some consumers that Google collects monopoly rents because it controls 78 percent of the search-advertising business. The FTC's rationale is that Google AdWords operates as an auction, so it can't set premium prices. But as Google's own AdWords blog points out, even if you were the only bidder on a given keyword, "the minimum CPC [cost per click] for a keyword is not related to the number of competitors one has for that keyword." Google has a minimum threshold you need to meet in order for your ads to be able to show. Therefore, by setting minimum bids, Google is clearly able to charge monopoly rents.

But it is in the area of data collection where Facebook and Google are going to offer the most resistance. The FTC has been completely silent on this subject, in contrast to the European Union. The EU has raised the issue of what degree of control individuals have over their digital data, ranging from pictures on Facebook to Web searches and surfing habits stored by Google. The EU wants to allow individuals the power to delete their data, which Internet companies would otherwise store and monetize through targeted ads. Given that Google stores your complete search history, your location history, your purchase data, your demographic profile, your calendar, and your contacts, it would seem only logical that the FTC should consider a default option that asks if you want to opt in to data sharing.

In addition, the FTC needs to investigate ad fraud. Following is the scope of the problem:

- As much as 50 percent of publisher traffic is bot activity – fake clicks from automated computing programs.
- Bots account for 11 percent of display-ad views and 23 percent of video-ad views.
- Digital advertising will take in $43.8 billion in 2017, $6.3 billion of which will be based on fraudulent activity.

So who pays for the $6.3 billion in fraudulent ad charges? First the advertisers pay it, then they pass the cost on to consumers. Of course for Google and other Internet ad exchanges, it's not their problem; it's built into their business model. But as with every part of this puzzle, the ability to fix the problem is fairly simple. If a firm such as White Ops can tell the difference between bot views and human views, so can Google. We at the Annenberg Innovation Lab discovered this when we published a report on ad exchanges that send excess ad inventory to pirate sites to "liquidate." Within months of being named as the worst offender, Google stopped most of the ads from landing on these sites.

3.

The idea of monopoly as the apotheosis of American capitalism goes to the heart of whether Tim Berners-Lee's vision of a "re-decentralized web" can actually be fulfilled. To my mind, Robert Bork's notion that the only stakeholder that matters in antitrust considerations is the consumer runs counter to every belief Americans have held about the role of big business, from the foundation of the nation to the election of Ronald Reagan. Writing in *Mother Jones,* Kevin Drum notes:

> If we want a more dynamic economy, we should quit protecting large, hulking incumbents. Our key interest should be ensuring plenty of competition, not engaging in pointless arguments about consumer welfare — a concept that's inherently fuzzy and far too easy to manipulate by smart corporate legal teams. To accomplish this, antitrust law needs to return to its older, cruder goal of simply keeping companies from getting too big and too dominant.

Clearly Thomas Jefferson and James Madison believed that controlling the threat of monopoly was so important to democracy that "freedom from monopoly" should be in the Bill of Rights. And as the power of business grew after the Industrial Revolution, the central battle of the early twentieth century was to break up monopoly trusts — a battle undertaken by Republicans Theodore Roosevelt and William Howard Taft as well as Democrat Woodrow Wilson.

Equally important, we discovered in the age of information that "natural monopolies" are often rational ways to achieve efficient networks. This picture of what the streets of New York looked like in 1895 shows why sometimes natural monopolies make sense.

Clearly the proliferation of telephone and telegraph networks, none of them interoperable, was a kind of tragedy of the commons. No one got any network effects – which made the networks themselves almost useless. The solution was for two companies, AT&T (then known as the Bell System) and Western Union, to consolidate the industries, buying up all the small operators and forming a natural monopoly. By the 1920s there was one phone company, one telegraph company, and only one communication wire into every home. But of course the government supervised these monopolies through the FCC. The Bell System had its rates regulated and was required to spend a fixed percentage of its monopoly profits on research and development that would be beneficial to the society at large.

In 1925 AT&T set up Bell Labs as a separate operating subsidiary with the mandate not only to develop the next generation of communications technology but also to conduct basic research in physics and other sciences. Over the course of the following fifty years, the transistor, the microchip, the solar cell, the microwave transmitter, the laser, and cellular telephony all were invented at Bell Labs, along with myriad other things we take for granted. Eight Nobel Prizes have been awarded to Bell Labs scientists. It was the most productive research facility in history; in fact most of the achievements of the digital age rest on its inventions. In the 1956 consent decree in which the Justice Department allowed AT&T to maintain its phone monopoly, the government extracted a huge concession. As explained by Jon Gertner in his book *The Idea Factory: Bell Labs and the Great Age of American Innovation*, "The phone company agreed to license its present and future U.S. patents to all American applicants 'with no limit as to time or the use to which they may be put.'" All the past patents were licensed royalty-free, and all the future patents were to be licensed for a small fee. The government made this deal because it considered the phone system a utility.

Could the AT&T remedies be applied to the natural monopoly called Google? First let's tackle the definition of "public utility." Have the services Google provides become so very, very central to the market that they have passed into the realm of utility? In finding for Google against the Authors Guild in an important case about Google's digitization of thirty thousand books, judge Denny Chin said the service offered "significant public benefits," even though Google was the only one profiting from it. So assuming one could argue that Google is a monopoly and needs to enter into a consent decree, would the Bell Labs model work?

If Google were required to license every patent it owns

for a nominal fee to any American company that asks for it, it would have to license its search algorithms, Android patents, self-driving car patents, smart-thermostat patents, advertising-exchange patents, Google Maps patents, Google Now patents, virtual-reality patents, and thousands of others. What is clear from the Bell Labs model is that such a solution actually benefits innovation in general. The availability of the Bell Labs transistor patents allowed the rise of Texas Instruments, Fairchild Semiconductor, and Intel. The cellular patents allowed Motorola to grow exponentially – only to be bought by Google for its patents. Bell Labs' satellite patents created a whole new industry with many players. And we owe the current boom in solar energy to the original Bell Labs solar-cell patents.

It would seem that such a licensing program would be totally in line with Google's stated "Don't be evil" corporate philosophy and could spawn the birth of many new firms. However, Google's use of patents has often crossed the line. They were censured by the FTC in 2012 over their efforts to block US imports of smartphones made by Microsoft and Apple by asserting that the devices, which rely on industry-standard technology, infringe patents owned by Google's Motorola Mobility unit. Bert Foer, president of the American Antitrust Institute, stated that "there is a tough emerging attitude by antitrust regulators who've recognized that the failure to honor standard essential patent commitments needs to be treated much more severely than in the past." Of course for such a remedy as I am proposing to work, some government agency would need to initiate a Google antitrust case. As I have pointed out, Google seems to have the most successful regulatory capture strategy of any company in decades. Whether that will change after the election of 2016 is anyone's guess, but it might just have to come from a state attorney general rather than the federal government, which seems to have totally bought into Robert Bork's theory of antitrust.

4.

There is one more aspect of the digital future that is potentially troubling: the lack of competition in the broadband business—the pipeline that brings the Internet to us. The proposed merger of the largest and second-largest broadband providers (Comcast and Time Warner Cable), which was blocked by the FCC, raised the specter of a single provider controlling 40 percent of all high-speed broadband in the United States. For the usually passive FCC, even that was a bridge too far. Comcast had made the argument that since each company operated a de facto monopoly in the individual cable markets they serve, their merger would not change the competitive environment for the individual consumer. While it is true that a broadband duopoly—one cable provider and one telecommunications company—is standard in most major markets, it should not be cause for celebration. This situation means that we have slower broadband service and pay higher prices for it than almost any developed country in the world. The US also lags behind twenty-one countries in the adoption of fiber-optic technology as this chart shows. The United Kingdom has the lowest fiber optic penetration of any developed country.

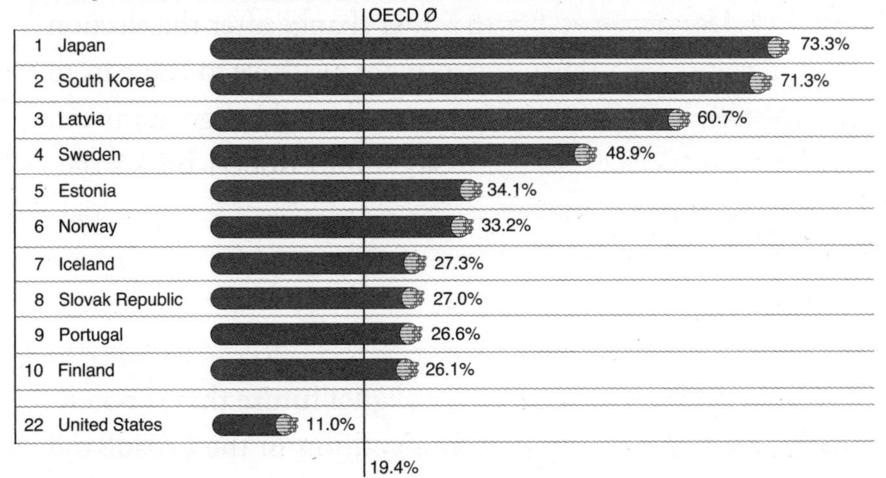

Since 2013, at the Annenberg Innovation Lab, we have had a chance to see what really fast broadband looks like. And no, we don't have to travel to Seoul, South Korea, to experience the future. We go to Chattanooga, Tennessee, where we can test applications at one gigabit per second over the EPB fiber network. EPB stands for the Electric Power Board of Chattanooga, a municipally owned utility. And now the company is deploying ten-gigabit-per-second service. The EPB's story points us toward a future in which we may no longer have to worry about a broadband duopoly.

A few years ago the folks at Volkswagen told the Chattanooga city fathers that they would like to build a high-tech auto plant in their city. There was only one problem: the city sits in the middle of Tornado Alley, and the electricity goes out several times a year during big storms. Since the plant was going to be highly roboticized, electrical outages would be particularly problematic. So the EPB promised to build a smart grid so that when a tree fell on the wires on Flynn Street, only Flynn Street would go dark, because the smart grid would route power around the trouble.

So they built the smart grid, Volkswagen built its plant, and the plant hasn't had any downtime. But once EPB had strung fiber-optic cable on every lamppost in town, it realized that each of these posts stood less than one hundred feet from a home to which the company could sell broadband service—and there were at least fifty thousand of these homes. Comcast, the local incumbent, tried to sue EPB and stop them in the Tennessee legislature. Once the EPB started advertising "the fastest broadband in Chattanooga," Comcast sent the company cease-and-desist letters, suggesting that its own network could provide service at one hundred megabits per second if there were only one person on the line. So the EPB improved its service to one gigabit per second, and the cease-and-desist letters stopped. Now the EPB is gaining market share for its $70 monthly plan, which gives consumers one-gig broadband speed plus TV service. Comcast is losing market share with its old-fashioned cable broadband product.

Why is this story important to our country's technological future? Because it's emblematic of what can happen when true competition is ignited in the broadband market. In early 2014 the new FCC chairman, Tom Wheeler, released a statement on "open Internet" rules in the wake of a federal court decision on network neutrality. In a section on enhancing competition, he wrote, "One obvious candidate for close examination [is] legal restrictions on the ability of cities and towns to offer broadband services to consumers in their communities." In the federal court case *Verizon v. FCC*, Judge Laurence Silberman suggested that the FCC's primary obligation was to promote competition and remove barriers to infrastructure investment. Those barriers have been constructed by cable-company and telecommunications-company incumbents and their lobbyists, who have persuaded about twenty state legislatures to pass bills restricting municipalities from entering into the broadband market. Judge Silberman described these laws as providing "an economic preference to a politically powerful constituency, a constituency that, as is true of typical rent seekers, wishes protection against market forces."

So the scrappy folks at EPB proposed to expand their fiber-optic service to the surrounding counties, but AT&T got its lobbyists at the Tennessee state legislature to pass a bill saying that municipally owned utilities could not compete with private firms such as AT&T and Comcast in the provision of broadband. Remarkably, the FCC agreed with the EPB and issued a preemption order, which barred the Tennessee legislature from blocking the EPB. And then of course the Tennessee attorney general sued the FCC. The FCC and EPB lost the first round in Federal District Court. I suspect the battle will go on for a while, because monopolists hate the idea that a small city could start its own network.

But as an observer from California who has visited Chattanooga seven times in five years, I can see how truly fast broadband can transform a town. Chattanooga was one of those southern cities that got screwed by globalization. The Brookings Institution reported, "With its extensive railroads and river access, Chattanooga was at one time the 'Dynamo of Dixie' — a bustling, midsized, industrial city in the heart of the South. By 1940, Chattanooga's population was centered around a vibrant downtown and it was one of the largest cities in the United States. Just 50 years later, however, it was in deep decline. Manufacturing jobs continued to leave." When I first started visiting the city, in 2010, months after EPB had deployed its fiber-optic technology to home networks, the downtown area was still filled with empty factory buildings. In 2016, an incredibly vibrant tech community has filled many of those vacant buildings with open-office spaces for start-ups. The music and film communities are vital, too, and there is a real downtown nightlife with bars and restaurants open late and filled with hipsters. You would think you were in Brooklyn, except the food and music are definitely southern. In 2012 at an outdoor music festival, the Annenberg Innovation Lab created a virtual country music duet over two thousand miles of fiber-optic cable. T Bone Burnett performed in my studio at USC, while Chuck Mead performed onstage in Chattanooga.

5.

What happened in Chattanooga gets to the heart of the idea of decentralization, the core of the original World Wide Web vision. Given the political paralysis in Washington, brought about by the power of money to control government, it is only right that productive experimentation is going to happen at the state and city levels. The Catholic Church has a concept called subsidiarity, which the *Oxford*

English Dictionary defines as "the idea that a central authority should have a subsidiary function, performing only those tasks which cannot be performed effectively at a more immediate or local level." Yuval Levin, in his book *The Fractured Republic: Renewing America's Social Contract in the Age of Individualism,* makes the case for subsidiarity.

> Honing an inclination to subsidiarity would offer us a way of thinking about solving problems together that begins in the neighborhood, in the church, in the school, in the community and builds up. It would mean a political system and a government better suited to meeting Americans where they are, better adapted to the range and variety of problems our country now confronts, better positioned to help us try solutions that arise in places as close to the problems as possible — and then to use our modern networked architecture of decentralized communication to teach and learn from others dealing with similar problems — and better able to revitalize our civic culture.

One of the fruits of this "think globally, act locally" strategy is found in the concept of the cooperative. A co-op is defined as a jointly owned enterprise engaging in the production or distribution of goods or the supplying of services, operated by its members for their mutual benefit. It seems to me that a co-op is the ideal way for producers of creative work to band together and get their content distributed at a fair price.

Just consider the history of the most famous producer co-op in the United States, Sunkist Growers, Inc. Sunkist is a marketing and distribution co-op for the four thousand growers of citrus fruit in California and Arizona. It is owned and managed by the growers themselves and had gross revenues of $1.1 billion in 2013. The co-op was formed in California in the 1890s to allow citrus growers to assert some collective power. Before the formation of the co-op, packinghouse owners, distributors, agents, and speculators — the middlemen — ruled, and the growers ranked a distant last in terms of exercising control over the industry. They were independent, small-scale farmers presiding over modest five-, ten-, and fifteen-acre groves without the organization and training to distribute their produce effectively. Their bargaining position got so bad that they were taking a loss on every crop, and the middlemen were earning high profits. In 1893 they formed the Southern California Fruit Growers Exchange. By regulating shipments and directing fruit where demand was the highest, the exchange made an immediate impact on the financial well-being of its growers. By the end of the first season growers realized an average net price of roughly $1 per box of oranges, a return far better than the twenty cents per box they had made the previous season. Within ten years the co-op had signed up about half of the growers in California. In 1908 the group registered the trademark "Sunkist" and later began stamping it on all its fruit, thus creating the first branded agricultural product. The co-op grew so strong that during the Depression, when prices plummeted, it was able to extend credit to its members so that very few of them had to sell their farms. When World War II arrived, the need for vitamin C–packed fruit juice to prevent scurvy led to many years of successful growth at Sunkist.

I cite the case of Sunkist because many of today's musicians and filmmakers are very much in the position of citrus growers in the 1890s. They are the key producers of their product, but they have no leverage and end up being an afterthought in the supply chain. The stunning thing about a zero-marginal-cost digital distribution system is that we forget that it could allow artists to run a nonprofit distribution cooperative and keep a far higher percentage of their revenues than they do now. YouTube takes 45 percent of the ad revenue on its site simply for running the infrastructure, without putting up production or marketing money. At worst the cost to run the infrastructure is around 5 percent, at current revenue levels, so the rest of the revenue is pure profit. What if artists ran a video and audio streaming site as a nonprofit cooperative (perhaps employing the technology in some of those free Google patents)? Let's assume they would let the co-op keep 10 percent of the revenue, either from ads or subscriptions, in order to run the infrastructure and have money for general marketing. The artist would take the remaining 90 percent. Then, as Netflix does, the co-op could rent cloud service from Amazon, Microsoft, or IBM on a worldwide basis. I realize this system might have to start in the music business, where the cost of production is fairly low and most artists have access to the digital tools required to make their own music. But when you realize that more than one thousand independently financed feature films are made each year, without distribution commitments, there is clearly going to be a need for a film distribution co-op as well.

Could the Sunkist model work in an artistic field? Well, it already has. Magnum Photos was formed by Robert

Capa, Henri Cartier-Bresson, and ten other freelance photographers at the end of World War II. Magnum's official history tells the story:

> It was founded as a co-operative in which the staff, including co-founders Maria Eisner and Rita Vandivert, would support rather than direct the photographers. Copyright would be held by the authors of the imagery, not by the magazines that published the work. This meant that a photographer could decide to cover a famine somewhere, publish the pictures in "Life" magazine, and the agency could then sell the photographs to magazines in other countries, such as Paris Match and Picture Post, giving the photographers the means to work on projects that particularly inspired them even without an assignment.

Seventy years later, Magnum is still thriving, allowing its artists to control their work and pursue their passions. This is a critical matter for creative people. Then there is Deca, which describes itself as "a global network of independent journalists, including Pulitzer, Polk, PEN, and National Magazine Award winners and finalists, who come together every year to bring a single, vital topic into sharp focus."

In Brooklyn, a new filmmakers' cooperative called SRSLY (short for "Seriously"), centered around female-identified filmmakers and feminist content, has begun making important work. Caroline Conrad, one of the founders, told *Brooklyn Magazine,* "One of our biggest motivations was to provide mentorship and guidance for young female filmmakers. Unfortunately, support is extremely rare. Women in film are often driven to see one another as competitors rather than collaborators." And in an article about the music collectives Odd Future and ASAP Mob, the *New York Times* wrote, "The longstanding principles of individual success, taken for granted by an older generation, are moving out of fashion. With digitally enabled peer-to-peer everything, from riding to working to living, emerging artists and entrepreneurs are creating new paradigms for coexisting."

Like Magnum Photos, these cooperatives do not ask artists to forgo other distribution outlets but rather enable them to take advantage of a series of distribution windows that could lead to much higher income, even if it means less for the middlemen. This model exists in the music business in the form of Bandcamp. The *New York Times* music critic Ben Ratliff explains:

> [Bandcamp] is known for its equitable treatment of artists, and [is] one of the greatest underground-culture bazaars of our time. From it, you can stream music to the extent each artist allows, or buy songs at a price set by the artist – which is sometimes "pay what you wish" – or order physical products from the site. The artist gets 85 percent. Always, the artist gets to know who's buying, without a third party in the way. There's also a social-media application on the site that lets the consumer know who else is buying and what else they've bought in the past. That is significant: You can triangulate your taste with other people, whom you don't know, but whom you might come to trust.

We can imagine a future in which artists would release first to their fans through a site like Bandcamp, taking in 85 percent of the subscription revenue. The second release, three weeks later, would be to Apple Music and Spotify Premium, from which the artist would probably get 70 percent of the subscription revenue. The third release, four weeks after that, would be to ad-supported streaming services such as Spotify and YouTube, from which, as I have shown, the artist's percentage would be much smaller. In this way, fans who want to hear the music immediately would pay more than casual fans who felt they could wait seven weeks for the record. Both Taylor Swift and Adele have used this windowing strategy with great success, but there is no reason most other artists couldn't do the same – except that when some record labels held back on releasing their big hits on ad-supported streaming services, both Spotify and Google (which hosts the Spotify service and seems to have some say in its direction) have threatened them with going to the antitrust authorities.

This raises the question of what role a traditional record company would play in a co-op ecosystem. My sense is that we would end up with a kind of parallel universe in which traditional record companies are responsible for artist development and risk aggregation, but only for certain kinds of artists. Others, those in co-ops, would end up like the members of Magnum – capable of producing their own work. They would run their own social media promotion and book their own tours. Big-name artists might have no need for a record company, and many new indie artists might have no desire to be attached to one. But in the middle, there are probably a lot of artists who would stay with traditional labels. In an ideal world, this kind of decentralized infrastructure of artist co-ops might bring back some of the regional distinctions that seemed so important when I was first on the road with Dylan and The Band. The music in San Antonio was then very different from the music in Austin. New Orleans was different from Memphis. Chicago R & B didn't sound anything like Detroit R & B. The sound of Los Angeles was different from the sound of San Francisco. Though there are regional hip-hop styles, regionalism doesn't exist today for most music. I hope it will return.

6.

Finally, in order to create a new American renaissance, we need a really good public media system. What I am imagining is a single well-funded TV, radio, and Internet service that integrates content and stories across all three platforms. The best current example I can point to is the BBC. Every British household pays a small yearly fee that finances the whole system. While the BBC is well-loved, it is regularly besieged by libertarian-influenced politicians who object to its competition to the commercial broadcasting services. In the United States, we could raise this money in several ways: a spectrum tax that broadcasters would pay for the use of public airwaves; a small tax on advertising revenue, which could easily pay for an advertising-free zone on public media; or selling much of the spectrum controlled by local PBS stations and putting the money in a production trust that would throw off enough income to pay for yearly production expenses.

The key to a vital public media system is the balance between local and global. My sense is that National Public Radio is doing a good job at this balance. Today NPR has 36.6 million listeners for its radio programming and 33.2 million viewers of its Web content. My local NPR station, KCRW, in Santa Monica, is a strong example of this local-global strategy. In addition to broadcasting national programs from NPR, American Public Media, and Public Radio International, KCRW produces four daily programs, eight weekly shows, and twenty-eight podcasts covering news, politics, entertainment, music, film, food, literature, design, architecture, and storytelling—all on an annual budget of $22 million. It has become the source of my continuing music education as well as the way I learn more about local food, art, and culture than I ever could from any other source, including the *Los Angeles Times*.

But while NPR is thriving, public television is a total mess. PBS has 350 stations, whereas a network like NBC has only 220 affiliates. This matters a lot, because the majority of the money that the government-funded Corporation for Public Broadcasting doles out every year goes to local stations, not to national programming. Not only is PBS starved for programming funds, there is also no innovation on the national programming schedule. PBS's anchor programs — *Masterpiece, Frontline, Nova,* and *PBS NewsHour* — have all been going for at least twenty-five years. Even *Sesame Street* has defected to HBO as a showcase for its first-run content. There has not been a successful new series in the last fifteen years.

PBS needs to radically shrink the number of stations it funds. Perhaps the model of having only a single public television station in every state, as there is in Vermont, points the way forward. Since 90 percent of TV viewers are getting the programming through cable or satellite, one could easily place repeater transmission towers around the state to reach the folks who are still using antennas. If PBS shrank the number of stations and used the money from spectrum sales and government grants for programming, we would have a genuine alternative to commercially sponsored TV in America. The strategy of investing in programming has worked in England: figures show that the commercial-free BBC networks had a weekly reach in April and June of 2015 encompassing 78.8 percent of the British TV audience, which spent an average of nine and a half hours per week on the networks. Needless to say, much of the BBC's programming investments end up as the best shows on PBS.

What intrigues me about the BBC model is that we need to break the grip of advertising on our media system. Advertisers have a natural bias against controversy—edgy art does not help sell sugar water. But the story of progress in art is all about edgy material. Imagine Picasso having to persuade an executive at Pernod to support his earliest cubist paintings. Or a Coca-Cola marketer in Chicago listening to Louis Armstrong's breakthrough "West End Blues" and wondering, "Will it help sell soda?" I think the growth of advertising in our time is symbolic of a deep crisis in capitalism. I have written earlier in this book of the great stagnation—the period since the early 1970s when median wages refused to rise even though productivity was rising dramatically. In the face of such wage stagnation, firms have to spend more money on marketing to get consumers to keep buying. As you wander down the supermarket aisle perusing endless varieties of detergent, all basically containing the same ingredients, the only distinguishing factor is the marketing pitch. Will Tide get my shirts cleaner than Surf? Behind most detergent brands are just two firms, Procter & Gamble and Unilever, each of which spends more than $8 billion per year on advertising.

As this stagnation — unbelievably — continues, the need for more intrusive "behavioral marketing" increases, and the Internet is the conduit for most of that — both on the demand side and the supply side. The supply is the constant surveillance of your data on billions of smartphones. The demand is advertisers' endless need to send messaging that recognizes the time of day, your mood, your device, and your geographic location. As early as the mid-1950s, the economist John Kenneth Galbraith suggested that there would come a time when "it can no longer be assumed that welfare is greater at an all-around higher level of production than at a lower one. . . . The higher level of production has, merely, a higher level of want creation necessitating a higher level of want satisfaction." Galbraith's assertion perfectly captures the goal of modern advertising: to create desire for products we didn't know we needed. The American middle class is in the same position as a hamster on a wheel: running faster and faster but making no progress in relation to its neighbors.

Galbraith could have never imagined the heights of want creation that Facebook could deliver when it's viewed on a mobile device we cannot ignore.

When I was at the monastery in Big Sur, I read a book by E. F. Schumacher called *Small Is Beautiful: Economics as if People Mattered.* One passage stood out for me as I was contemplating whether our smartphones were liberating us or just addicting us to more consumption:

> [The modern Western economist] is used to measuring the "standard of living" by the amount of annual consumption, assuming all the time that a man who consumes more is "better off" than a man who consumes less. A Buddhist economist would consider this approach excessively irrational: since consumption is merely a means to human well-being, the aim should be to obtain the maximum of well-being with the minimum of consumption.

But the business models of Google, Amazon, and Facebook are all built around the assumption that they hold the secret formula to continually arousing our desires. This may not be good for America, but it's good for Google, Amazon, and Facebook.

Afterword

I understand that I could be accused of elitism by contrasting Francis Ford Coppola with PewDiePie. I plead guilty, but in part because I proudly spent the first thirty years of my career producing music, movies, and TV, I am eager to see those traditions continue. I also want to consume this kind of media — I don't want to miss out on understanding what artists have to say about our culture. The question of how art lasts matters to me, and I think it should matter to you. I know that amazing music, film, journalism, and TV are being created by my children's peers. In fact, my own daughter is making wonderful movies such as *Beasts of No Nation* and *The Kids Are All Right*. But I also know the incredible struggle she goes through to get each movie financed. So if the digital revolution has devalued the role of the creative artist in our society, then we need to do more than play around the edges. So often when I have given a speech on this topic to musicians, book publishers, moviemakers, and authors, I get what I have come to call a Stockholm syndrome response — "Isn't this just the way things are? Don't we have to resign ourselves to working with Google, Facebook, and Amazon?" The audiences seem to parrot the techno-determinism that is, after all, just one way of understanding the problem.

I think big changes could happen if we approach the problem of the monopolization of the Internet with honesty, a sense of history, and a determination to protect what we all agree is important: our cultural inheritance. We all need the access to information the Internet provides, but we need to be able to share information about ourselves with our friends without unwittingly supporting a corporation's profits. Facebook and Google must be willing to alter their business model to protect our privacy and help thousands of artists create a sustainable culture for the centuries, not just make a few software designers billionaires. We also must understand that the men who run Google, Facebook, and Amazon are just at the beginning of a long project to change our world. Yuval Noah Harari calls their project Dataism:

> Dataists further believe that given enough biometric data and computing power, this all-encompassing system could understand humans much better than we understand ourselves. Once that happens, humans will lose their authority, and humanist practices such as democratic elections will become as obsolete as rain dances and flint knives.

We need to confront this techno-determinism now — with real solutions — before it is too late. I undertake the pursuit of these solutions with both optimism and humility. Optimism because I believe in the power of rock and roll, books, and movies to upset the world. As the writer Toni Morrison observed, "The history of art, whether it's in music or written or what have you, has always been bloody, because dictators and people in office and people who want to control and deceive know *exactly* the people who will disturb their plans. And those people are artists. They're the ones that sing the truth. And that is something that society has got to protect." I know that brave and passionate art is worth protecting and is more than just click bait for global advertising monopolies. I know it can change lives.

Certainly witnessing Bob Dylan go electric at the Newport Folk Festival in 1965 turned this Princeton freshman who had previously been intent on being a lawyer into a passionate follower of the rock-and-roll circus – one who managed to make a good living from the entertainment business. My optimism also showed itself in 1996 when I helped found one of the first streaming video-on-demand services. Anyone crazy enough to found a service that needed broadband in 1996 had to be an optimist. My optimism led to humility, because the diffusion of broadband was much slower than I thought; I know that predicting the future is a humble man's game.

I think many in my generation have a utopian impulse (which is, it should be observed, different from idealism), but it is slipping away like a short-term memory. I feel I need to quote Dr. King again, who said the night before he was assassinated, "I may not get there with you, but I believe in the promised land." My generation knew that the road toward a better society would be long, but we hoped our children's children might live in that land, even if we weren't able to get there with them. It may take even longer than we imagined to rebuild a sustainable culture. If I were to predict the future, I would hope to see Tim Berners-Lee's dream of a "re-decentralized" Internet, one that's much less dependent on surveillance marketing and that allows creative artists to take advantage of the zero-marginal-cost economics of the Web in a series of nonprofit distribution cooperatives. I have no illusion that the existing business structures of cultural marketing will go away, but my hope is that we can build a parallel structure that will benefit all creators. The only way this will happen is that in Peter Thiel's "deadly race between politics and technology," the people's voice (politics) will have to win. Google, Amazon, and Facebook may seem like benevolent plutocrats, but the time for plutocracy is over.

Acknowledgments

This book grew out of a speech I gave at the Aspen Ideas Festival in the summer of 2015. Thanks to Walter Isaacson, Charlie Firestone, Kitty Boone, and John Seely Brown for making that possible. My agent, Simon Lipskar, encouraged me to make it into a book and introduced me to my editor, Vanessa Mobley at Little, Brown and Company. It is Vanessa who truly helped me realize the vision of the book, and for that I am ever grateful.

Much of the development of the ideas in the book came during my tenure as a professor at the Annenberg School for Communication and Journalism at the University of Southern California. I was fortunate to work under two extraordinary deans, Geoffrey Cowan and Ernest Wilson. I was also fortunate that I arrived at USC just as Professor Manuel Castells became a university professor at Annenberg. Manuel mentored me and got me to understand the power of networks. His work has informed me ever since. It was Dean Wilson who had the courage to support the Annenberg Innovation Lab that I have directed for the past six years. At the Lab I was surrounded by the amazing intellects of fellows like Henry Jenkins, Francois Bar, Gabriel Kahn, Elizabeth Currid-Halkett, and Robert Hernandez. The staff who run the Lab — Erin Reilly, Rachelle Meredith, Sophie Madej, and Aninoy Mahapatra — helped make every project an adventure and taught me more than I could ever teach them.

My experience in culture production has always been in the film, TV, and music businesses, so the guidance into the intricacies of book production provided by Ben Allen and Barbara Clark were invaluable. The same goes for Sabrina Callahan and Lauren Passell.

In the battle for artist rights I have taken inspiration from my colleagues T Bone Burnett, David Lowery, Brian McNelis, Chris Castle, and Jeffrey Boxer. We all know it is a long fight against very powerful forces, but no one is giving in.

And finally a note of deep gratitude to my wife, Maggie, and our children — Daniela, Nick, and Blythe. Your lives have been an inspiration to me. Your love has anchored me.

Notes

Introduction

Holman Jenkins, "Technology=Salvation," *Wall Street Journal,* October 9, 2010, www.wsj.com/articles/SB10001424052748704696304575537882643165738.

Craig Silverman, "Viral Fake Election News Outperformed Real News on Facebook in Final Months of the U.S. Election," *Buzzfeed,* November 16, 2016, www.buzzfeed.com/craigsilverman/viral-fake-election-news-outperformed-real-news-on-facebook.

My calculation on the $50 billion a year reallocation of revenue from content creators to tech platforms goes as follows: According to the Newspaper Association of America, newspaper ad revenues have fallen from $65.8 billion in 2000 to $23.6 billion in 2014. So that's a $42.2 billion shortfall. According to the Record Industry Association of America, recorded music revenues have fallen from $19.8 billion in 2000 to $7.2 billion in 2014. So that is another $12.6 billion shortfall. According to the Digital Entertainment Group, video home-entertainment revenues were $24.2 billion in 2006 and $18 billion in 2014. So that is another $6 billion per year shortfall. In total, that is $61 billion a year less to content creators and that doesn't even count the book business, whose revenues have remained essentially flat because the juvenile business has almost made up for the 23 percent decline since 2007 in nonfiction sales, and the 37 percent decline in adult fiction sales. These statistics come from Nielsen Bookscan and include ebook sales.

Perfect Search Media did a report and concluded that "Entertainment products and services are the most searched online products and services in Google." "Google's Most Searched Online Products and Services-June 2013," Slideshare, December 5, 2013, www.slideshare.net/PerfectSearchMediaDesign/perfect-search-mostsearchedonlinegoogle.

Barry Lynn and Phillip Longman, "Populism with a Brain," *Washington Monthly,* August 2016, washingtonmonthly.com/magazine/junejulyaug-2016/populism-with-a-brain/.

Martin Luther King's final sermon at the National Cathedral in Washington, DC, was preached on March 31, 1968. Martin Luther King Jr., "Remaining Awake Through a Great Revolution," The King Institute, kinginstitute.stanford.edu/king-papers/publications/knock-midnight-inspiration-great-sermons-reverend-martin-luther-king-jr-10.

Leon Wieseltier, "Among the Disrupted," *New York Times,* January 7, 2015, www.nytimes.com/2015/01/18/books/review/among-the-disrupted.html.

Carole Cadwalladr, "Google, Democracy and the Truth About Internet Search," *Guardian,* December 4, 2016, www.theguardian.com/technology/2016/dec/04/google-democracy-truth-internet-search-facebook.

Dan Kaminsky's speech at the 2015 Black Hat Conference cited the NTIA report for his contention that 50 percent of Internet users are worried about security. National Telecommunications and Information Administration, "Lack of Trust in Internet Privacy and Security May Deter Economic and Other Online Activities," May 13, 2016, www.ntia.doc.gov/blog/2016/lack-trust-internet-privacy-and-security-may-deter-economic-and-other-online-activities.

Jacob Silverman, "Just How Smart Do You Want Your Blender To Be?" *New York Times,* June 19, 2016, www.nytimes.com/2016/06/19/magazine/just-how-smart-do-you-want-your-blender-to-be.html.

Chapter One: The Great Disruption

Jorge Guzman and Scott Stern, "The State of American Entrepreneurship: New Estimates of the Quantity and Quality of Entrepreneurship in 15 US States, 1988–2014," March 2016, static1

.squarespace.com/static/53d52829e4b0d9e21c9a6940/t/56d9a05545bf217588498535/1457102936611/Guzman+Stern+—+State+of+American+Entrepreneurship+FINAL.pdf.

David Nasaw, *Andrew Carnegie* (New York: Penguin Books, 2007).

OECD Economics Department, "Policy Note No. 24: Shifting Gear: Policy Challenges for the Next 50 Years," June 2014.

What is critical to note is that although the digital economy currently represents only 22.5 percent of the global economy (according to Accenture), it will represent 25 percent by 2020 and keep growing at a far faster rate than the nondigital economy.

Chapter Two: Levon's Story

Though most of the story of Levon Helm and The Band comes from my own experience with them, there are two good books on that period in Woodstock by British journalist Barney Hoskins: *Across the Great Divide: The Band & America* (New York: Hal Leonard, 2006) covers most of their career. *Small Town Talk: Bob Dylan, The Band, Van Morrison, Janis Joplin, Jimi Hendrix and Friends in the Wild Years of Woodstock* (New York: Da Capo Press, 2016) has a much wider view of the whole scene in late 1960s Woodstock.

Chris Anderson's book from 2008, *The Long Tail: Why the Future of Business Is Selling Less of More* (New York: Hachette, 2008) is still fairly controversial. I believe he underestimated the power of search engines to push only the most popular content to the top.

The best recording ever of Ray Charles is *Ray Charles Live* from a 1959 Atlanta stadium show with a big band. This is where you can find the inexorably cathartic and impossibly slow "Drown in My Own Tears."

Chapter Three: Tech's Counterculture Roots

Although I never got to meet Doug Engelbart, I was fortunate enough to have spent time with some of the founders of the Internet, including Vint Cerf, Tim Berners-Lee, and most especially

John Seely Brown, who has been a mentor to me for the past seven years.

Thierry Bardini, *Bootstrapping: Douglas Engelbart, Coevolution, and the Origins of Personal Computing* (Palo Alto: Stanford University Press, 2000).

Fred Turner, *From Counterculture to Cyberculture* (Chicago: University of Chicago Press, 2008), and John Markoff, *What the Dormouse Said* (New York: Viking, 2005), are both wonderful resources around the story of the early Internet.

Lee Vinsel and Andrew Russell, "Hail the Maintainers," *Aeon,* April 7, 2016, aeon.co/essays/innovation-is-overvalued-maintenance-often-matters-more.

Chapter Four: The Libertarian Counterinsurgency

Peter A. Thiel and David O. Sachs, *The Diversity Myth* (Oakland: The Independent Institute, 1998).

Peter A. Thiel and Blake Masters, *Zero to One: Notes on Startups, or How to Build the Future* (New York: Crown Business, 2014).

Jennifer Burns, *Goddess of the Market: Ayn Rand and the American Right* (London: Oxford University Press, 2009).

Anne Conover Heller, *Ayn Rand and the World She Made* (New York: Nan A. Talese, 2009).

Ayn Rand, *The Fountainhead* (New York: Bobbs-Merrill, 1943).

Ayn Rand, *Atlas Shrugged* (New York: Random House, 1957).

Peter Thiel, "The Education of a Libertarian," Cato Institute, April 2009, www.cato-unbound.org/2009/04/13/peter-thiel/education-libertarian.

Hans Hermann-Hoppe, *Democracy – The God That Failed* (Newark: Transaction, 2001).

Greg Satell, "Peter Thiel's Four Rules for Creating a Great Business," *Forbes,* October 3, 2014, www.forbes.com/sites/gregsatell/2014/10/03/peter-thiels-4-rules-for-creating-a-great-business/2/#2ea53ac12804.

Brad Stone, *The Everything Store: Jeff Bezos and the Age of Amazon* (New York: Little, Brown and Company, 2013).

Chapter Five: Digital Destruction

The film *The Social Network*, written by Aaron Sorkin and directed by David Fincher, is a remarkably honest telling of the birth of Facebook.

David Kirkpatrick, *The Facebook Effect* (New York: Simon and Schuster, 2010).

David Kirkpatrick, "With a Little Help From His Friends," *Vanity Fair,* September 6, 2010, www.vanityfair.com/culture/2010/10/sean-parker-201010.

Samantha Krukowski (ed.), *Playa Dust: Collected Stories from Burning Man* (San Francisco: Black Dog, 2014).

Ken Auletta, *Googled: The End of the World as We Know It* (New York: Penguin, 2009).

Most of the email quotes from YouTube come from depositions in the *Viacom International, Inc. vs. YouTube, Inc.* lawsuit.

Austin Carr, "Reddit Co-Founder, The Band's Ex-Tour Manager Debate SOPA, Anti-Piracy and Levon Helm's Legacy," *Fast Company,* April 19, 2012, www.fastcompany.com/1834779/reddit-cofounder-bands-ex-tour-manager-debate-sopa-antipiracy-and-levon-helms-legacy-video.

Kurt Andersen, "You Say You Want a Devolution?" *Vanity Fair,* January 2012.

Chapter Six: Monopoly in the Digital Age

Much of the research on Robert Bork came from an interview with former secretary of labor Robert Reich, who was both a student and a research assistant to Bork in the 1970s.

Robert Bork, *Antitrust Paradox* (New York: Basic Books, 1978).

Barry Lynn, *Cornered: The New Monopoly Capitalism and the Economics of Destruction* (New York: John Wiley and Sons, 2010).

Lee Epstein, William Landes, and Richard Posner, "How Business Fares in the Supreme Court," *University of Minnesota Law Review,* vol. 97, no. 1, www.minnesotalawreview.org/articles/volume-97-lead-piece-business-fares-supreme-court/.

Peter Orszag and Jason Furman, *A Firm-Level Perspective on the Roll of Rents in the Rise in Inequality,* presentation at "A Just

Society" Centennial Event in Honor of Joseph Stiglitz, Columbia University, October 16, 2015, www.whitehouse.gov/sites/default/files/page/files/20151016_firm_level_perspective_on_role_of_rents_in_inequality.pdf.

Martin Gilens and Benjamin I. Page, "Testing Theories of American Politics: Elites, Interest Groups, and Average Citizens," *Perspectives on Politics,* vol. 12, no. 3 (September 2014), scholar.princeton.edu/sites/default/files/mgilens/files/gilens_and_page_2014_-testing_theories_of_american_politics.doc.pdf.

Chapter Seven: Google's Regulatory Capture

Scott Cleland's *Precursor* blog at www.precursorblog.com is a valuable resource for anyone studying Google. Cleland has pointed out that Google, Facebook, Amazon, and Apple are about to engage in a new war for our attention, using voice-activated "digital assistants" placed around the home to perceive our desires. Google CEO Sundar Pichai told attendees of their I/O developer conference in May 2016 that "Our ability to do conversational understanding is far ahead of what other assistants can do. We're an order of magnitude ahead of everyone else." The future is here, where not just your Android smartphone or your Google Home assistant, but your Nest thermostat and security cam and your Google TV set can report your every movement and conversation back to Google servers—all to better sell advertising and products. We continue to surrender more of our private lives while believing in the myth of convenience donated for free by benign corporations.

The Google Transparency Project: googletransparencyproject.org.

Adam Pasick and Tim Fernholz, "The Stealthy Eric Schmidt–Backed Startup That's Working to Put Hillary Clinton in the White House," *Quartz,* October 9, 2015, qz.com/520652/groundwork-eric-schmidt-startup-working-for-hillary-clinton-campaign/.

Edmund Morris, *The Rise of Theodore Roosevelt* (New York: Random House, 2010).

James Lardner, "The Instant Gratification Project," *Business 2.0,* December 2001.

Chapter Eight: The Social Media Revolution

Harry McCracken, "Inside Mark Zuckerberg's Bold Plan for the Future of Facebook, *Fast Company,* November 16, 2015, www.fastcompany.com/3052885/mark-zuckerberg-facebook.

Daniel Hunt, "The Influence of Computer-Mediated Communication Apprehension on Motives for Facebook Use," *Journal of Broadcasting & Electronic Media,* vol. 56, no. 2 (June 2012).

David Kravets, "Facebook's $9.5 Million Beacon Settlement Approved," *Wired,* September 21, 2012, www.wired.com/2012/09/beacon-settlement-approved/.

Julia Angwin, *Dragnet Nation: A Quest for Privacy, Security, and Freedom in a World of Relentless Surveillance* (New York: Times Books, 2015).

Hazel Markus and Paula Nurius, "Possible Selves," *American Psychologist,* vol. 41, no. 9 (September 1986), psycnet.apa.org/index.cfm?fa=buy.optionToBuy&id=1987-01154-001.

The best work on Facebook and Google cooperation with the NSA PRISM program is a series of articles by Glenn Greenwald in the *Guardian* during the months of June and July 2013.

Kevin Cahill, "PRISM and the Law: The State of Play, August 2016," *Computer Weekly,* August 15, 2016, www.computerweekly.com/opinion/Prism-and-the-law-The-state-of-play-in-August-2016.

Bob Garfield, *The Chaos Scenario* (New York: Stielstra, 2009). Garfield also has a weekly radio show on NPR called *On the Media.*

Ben Elgin, Michael Riley, David Kocieniewski, and Joshua Brustein, "How Much of Your Audience Is Fake?" *Bloomberg Businessweek,* October 2015, www.bloomberg.com/features/2015-click-fraud/.

Ben Thompson's blog *Stratechery* is a must-read, stratechery.com.

George Packer, "No Death, No Taxes," *New Yorker,* November 28, 2011, www.newyorker.com/magazine/2011/11/28/no-death-no-taxes.

Tad Friend, "Tomorrow's Advance Man," *New Yorker,* May 18, 2015, www.newyorker.com/magazine/2015/05/18/tomorrows-advance-man.

Chapter Nine: Pirates of the Internet

Charles Graeber, "Inside the Mansion and Mind of the Net's Most Wanted Man," *Wired,* October 2012.

Google and PRS for Music commissioned report, "The Six Business Models for Copyright Infringement," June 27, 2012, www.prsformusic.com/aboutus/policyandresearch/researchandeconomics/Documents/TheSixBusinessModelsofCopyrightInfringement.pdf.

MUSO, "Global Music Piracy Insight Report 2016," July 2016, www.muso.com/market-analytics-global-music-insight-report-2016/.

USC Annenberg Innovation Lab, "Advertising Transparency Report," fifth edition, June 12, 2013, www.annenberglab.com/projects/ad-piracy-report-0.

J. M. Berger and Jonathon Morgan, "The ISIS Twitter Census," Brookings Project on U.S. Relations with the Islamic World, Analysis Paper No. 20, March 2015, www.brookings.edu/wp-content/uploads/2016/06/isis_twitter_census_berger_morgan.pdf.

Gareth Owen and Nick Savage, "The Tor Dark Net," Global Commission on Internet Governance, Paper Series No. 20, September 2015, www.cigionline.org/sites/default/files/no20_0.pdf.

Charlie Warzel, "'A Honeypot for Assholes': Inside Twitter's 10-Year Failure to Stop Harassment," *Buzzfeed,* August 11, 2016, www.buzzfeed.com/charliewarzel/a-honeypot-for-assholes-inside-twitters-10-year-failure-to-s.

Chapter Ten: Libertarians and the 1 Percent

Jane Mayer, *Dark Money: The Hidden History of the Billionaires Behind the Rise of the Radical Right* (New York: Doubleday, 2016). This is essential reading for anyone who wants to understand what happened to our democracy.

Robert McChesney, *Digital Disconnect: How Capitalism Is Turning the Internet Against Democracy* (New York: New Press, 2015). This is a really good primer from a media activist and scholar who has been fighting media monopoly for twenty-five years.

Robert Gordon, *The Rise and Fall of American Growth* (Princeton: Princeton University Press, 2016). This is the great antidote to digital utopians. Gordon's well-researched text shows that the Internet has not been the productivity bounty that was promised.

Sara C. Kingsley, Mary L. Gray, and Siddharth Suri, "Monopsony and the Crowd: Labor for Lemons?" Oxford Internet Institute, August 2014, ipp.oii.ox.ac.uk/2014/programme-2014/track-a/labour/sara-kingsley-mary-gray-monopsony-and.

Oxford University's Martin Programme on Technology and Employment is a critical resource for information on automation and the future of work. It can be found at www.oxfordmartin.ox.ac.uk/news/201501_Technology_Employment.

Andrew Gumbel, "San Francisco's Guerrilla Protest and Google Buses Swells into Revolt," *Guardian,* January 25, 2014, www.theguardian.com/world/2014/jan/25/google-bus-protest-swells-to-revolt-san-francisco.

Tom Perkins, "Progressive Kristallnacht Coming?" Letter to the Editor, *Wall Street Journal,* January 24, 2014, www.wsj.com/news/articles/SB10001424052702304549504579316913982034286.

David Graeber, *The Utopia of Rules: On Technology, Stupidity, and the Secret Joys of Bureaucracy* (London: Melville House, 2015). This is a funny, biting chronicle of the world of "bullshit jobs."

Chapter Eleven: What It Means to Be Human

Nir Eyal, *Hooked: How to Build Habit-Forming Products* (New York: Portfolio, 2014). Eyal takes B. F. Skinner's conditioning model to its logical conclusion.

Donald Trump's fake Twitter followers can be found at fakers.statuspeople.com/realdonaldtrump.

The best coverage of the Gamergate controversy has come from Caitlin Dewey of the *Washington Post.* The amount of bad information and gossip on this subject is astonishing.

David Auerbach, "Letter to a Young Male Gamer," *Slate,* August 27, 2014, www.slate.com/articles/technology/bitwise/2014/08/zoe_quinn_harassment_a_letter_to_a_young_male_gamer.html.

Michael Perilloux's writing on Neoreaction can be found at www.socialmatter.net.

Wael Ghonim's Ted Talk, "Inside the Egyptian Revolution," was filmed in March of 2011, www.ted.com/talks/wael_ghonim_inside_the_egyptian_revolution.

Pico Iyer, *The Art of Stillness* (New York: Simon and Schuster/TED, 2014). This is a slim volume, but full of wisdom. It was his suggestion that sent me to the New Camaldoli Hermitage in Big Sur, California.

Daniel Bell, *The Cultural Contradictions of Capitalism* (New York: Basic Books, 1976). First published in 1976, it is somewhat dated, but many of Bell's core findings still seem to apply to our current culture.

Jacques Barzun, *From Dawn to Decadence* (New York: Harper Collins, 2000). This is one of the truly great cultural histories.

John Seabrook, *The Song Machine: Inside the Hit Factory* (New York: W. W. Norton, 2016). This is a wonderful book on the modern pop-music business.

Frans de Waal, "How Bad Biology is Killing the Economy," *Evonomics,* March 2016, evonomics.com/how-bad-biology-is-killing-the-economy/.

Henry Jenkins, *Convergence Culture: Where Old and New Media Collide* (New York: NYU Press, 2006). Jenkins and I don't always agree, but he has been a constant source of inspiration and guidance for me at the Annenberg Innovation Lab.

Neil Postman, *Amusing Ourselves to Death* (New York: Penguin, 1985). This was a pre-Internet look at the role of popular culture in pacifying the American public. His belief that Aldous Huxley's vision of the future was correct is more true than ever.

Mark Grief, *The Age of the Crisis of Man: Thought and Fiction in America, 1933–1973* (Princeton: Princeton University Press, 2015).

Chapter Twelve: The Digital Renaissance

Christopher Moyer, "How Google's AlphaGo Beat Lee Sedol, a Go World Champion," *Atlantic,* March 28, 2016, www.theatlantic.com/technology/archive/2016/03/the-invisible-opponent/475611/.

Lawrence Summers and J. Bradford DeLong, "The 'New Economy': Background, Historical Perspective, Questions, and Speculations," Federal Reserve Bank of Kansas City, August 2001, www.kansascityfed.org/publicat/sympos/2001/papers/S02delo.pdf.

The Warner Music Group filing to the Registrar of Copyrights can be found at www.regulations.gov/document?D=COLC-2015-0013-86022.

Number of ad blockers downloaded worldwide: www.statista.com/statistics/435252/adblock-users-worldwide/.

National Association of Advertisers and White Ops, "The Bot Baseline Report," September 30, 2015, www.whiteops.com/bot-baseline.

Jon Gertner, *The Idea Factory: Bell Labs and the Great Age of American Innovation* (New York: Penguin, 2012).

John Curl, *For All the People: Uncovering the Hidden History of Cooperation, Cooperative Movements, and Communalism in America* (New York: PM Press, 2012).

Russell Miller, *Magnum: Fifty Years on the Frontline of History* (New York: Grove, 1998).

E. F. Schumacher, *Small Is Beautiful: Economics as if People Mattered* (New York: Harper, 1973).

Yuval Levin, *Fractured Republic: Renewing America's Social Contract in the Age of Individualism* (New York: Basic Books, 2016).

Toni Morrison, Ta-Nehisi Coates, and Sonia Sanchez, "Art is Dangerous," *VOX,* June 17, 2016, www.vox.com/2016/6/17/11955704/ta-nehisi-coates-toni-morrison-sonia-sanchez-in-conversation.

Yuval Noah Harari, "Big Data, Google, and the End of Free Will," *Financial Times,* August 26, 2016, www.ft.com/content/50bb4830-6a4c-11e6-ae5b-a7cc5dd5a28c.

Index

Index

Index

Index

Index

Index